ELECTRICAL SAFETY
AND THE LAW

A Guide to Compliance

Ken Oldham Smith

CEng, MIEE, MCIBSE

Fourth Edition

Revised by

John M. Madden

MSc, BSc, CEng, FIEE, MIOSH

Blackwell
Science

© 1990, 1993, 1997 Ken Oldham Smith
© 2002 Ken Oldham Smith and Blackwell Science
Ltd

Blackwell Science Ltd, a Blackwell Publishing
Company
Editorial Offices:
Osney Mead, Oxford OX2 0EL, UK
  Tel: +44 (0)1865 206206
Blackwell Science, Inc., 350 Main Street,
Malden, MA 02148-5018, USA
  Tel: +1 781 388 8250
Iowa State Press, a Blackwell Publishing
Company, 2121 State Avenue, Ames, Iowa
50014-8300, USA
  Tel: +1 515 292 0140
Blackwell Publishing Asia Pty, 550 Swanston Street,
Carlton South, Melbourne, Victoria 3053, Australia
  Tel: +61 (0)3 9347 0300
Blackwell Wissenschafts Verlag,
Kurfürstendamm 57, 10707 Berlin, Germany
  Tel: +49 (0)30 32 79 060

First Edition published 1990
Reprinted 1991 (twice), 1992
Second Edition published 1993
Reprinted 1993, 1994
Third Edition published 1997
Reprinted 2000
Fourth Edition published 2002

Library of Congress
Cataloging-in-Publication Data
is available

ISBN 0-632-06001-8

A catalogue record for this title is available from
the British Library

Set in 10 on 13pt Times
by DP Photosetting, Aylesbury, Bucks
Printed and bound in Great Britain by
TJ International Ltd, Padstow, Cornwall

For further information on
Blackwell Science, visit our website:
www.blackwell-science.com

ELECTRICAL SAFETY
AND THE LAW

Contents

Preface to the Fourth Edition

I first read Ken Oldham Smith's book in 1991 when I was researching the topic of electrical safety before attending a recruitment interview. I was impressed with the book and appreciated the quality of the information and advice that it contained. I was therefore honoured, some ten years later, to be given the opportunity to edit the book for its fourth edition, whilst being somewhat nervous about my ability to match the standards set by Ken.

In tackling this, I have set out to bring the references to legislation, standards and guidance material up to date. I have also worked to introduce two new chapters on topics that are of considerable relevance to electrical engineers working in the safety field. The first of these topics is safety-related electrotechnical control systems, and I have tried to explain the issues concerning the safety integrity of such systems, including those that incorporate programmable elements. Whereas the link to the more traditional electrical safety topics is slightly tenuous, this is a subject area that many engineers, technicians and electricians have to tackle, so having an awareness of the safety issues is especially important.

The second new topic is that of competence. Given that most accidents occur because of human error, or safety failures in systems of work, the management of competence is of fundamental importance.

These two subjects, of course, each merit a book in themselves, but I hope that I have encapsulated the main points in the new chapters.

The views and opinions expressed in the book are entirely my own and are based on my experience as an electrical engineer. I hope that they are informative and maintain the standards set in the first three editions of the book.

Finally, may I recognise with due appreciation the forbearance of my wife, Linda, who has had to put up with strewn papers and associated untidiness for what must have seemed an interminable period of time. Her support has been invaluable.

John Madden
Dunfermline, Fife
January 2002

Acknowledgement

The text of the Electricity at Work Regulations 1989 that is reproduced in Chapter 6 is subject to Crown copyright and is reproduced in accordance with the rules set out on the HMSO website www.hmso.gov.uk.

List of Abbreviations

ACB	air circuit breaker
ACOP	Approved Code of Practice
ALARP	as low as reasonably practicable
ASTA	Association of Short Circuit Testing Authorities
BASEC	British Approvals Service for Electrical Cables
BASEEFA	British Approvals Service for Electrical Equipment in Flammable Atmospheres
BEAB	British Electrotechnical Approvals Board
BSEN	British Standard Euronorm
BSI	British Standards Institution
CASS	conformity assessment of safety-related systems
CDM	Construction (Design and Management)
CEN	European Committee for Standardisation
CENELEC	European Committee for Electrotechnical Standardisation
CIBSE	Chartered Institution of Building Services Engineers
CNE	combined neutral/earth
CPC	circuit protective conductor
CTE	centre tapped to earth
DTI	Department of Trade and Industry
EC	European Community
ECA	Electrical Contractors' Association
EEBADS	earthed equipotential bonding and automatic disconnection of supply
EECS	Electrical Equipment Certification Service
EEMUA	Electrical Equipment Manufacturers and Users Association
EHO	environmental health officers
EMC	electromagnetic compatibility
EN	Euronorm
EOA	extension outlet assembly
ESPE	electrosensitive protective equipment
EU	European Union

FELV   functional extra-low voltage
HBC   high breaking capacity
HSC   Health and Safety Commission
HSE   Health and Safety Executive
HV   high voltage
IEC   International Electrotechnical Commission
IEE   Institution of Electrical Engineers
IP   ingress protection
ISA   incoming supply assembly
ISDA   incoming supply and main distribution assembly
LEL   lower explosive limit
LS   limit switch
LV   low voltage
MAG   metal active gas
MCB   miniature circuit breaker
MCCB   moulded case circuit breaker
MDA   main distribution assembly
MIG   metal inert gas
MIMS   mineral-insulated, metal-sheathed
MVA   million volt amperes
NICEIC National Inspection Council for Electrical Installation
        Contracting
NJUG   National Joint Utilities Group
OCB   oil filled circuit breaker
PE   protective earthing
PELV   protective extra-low voltage
PEN   combined neutral and protective conductor
PLC   programmable logic controller
PME   protective multiple earthing
PPE   personal protective equipment
PUWER   Provision and Use of Work Equipment Regulations
RCBO   residual current breaker with overcurrent device
RCCB   residual current circuit breaker
RCD   residual current device
RF   radio frequency
r.m.s. root mean square
RMU   ring main unit
SELECT Electrical Contractors' Association of Scotland
SELV   safety extra-low voltage
SF_6 sulphur hexafluoride
SI   Statutory Instrument
SIL   safety integrity level

SNE	separate neutral and earth
SOA	socket outlet assembly
TA	transformer assembly
TIG	tungsten inert gas
TRS	tough rubber sheathed
UEL	upper explosive limit
VRI	vulcanised-rubber-insulated

Chapter 1

The Hazards and Risks from Electricity

INTRODUCTION

The principal purpose of the legislation containing requirements for electrical safety is to reduce the number of injuries arising from the use of electricity. It is therefore important for both safety practitioners and people who work with electricity to have a good understanding of the type of injuries that can occur. This chapter sets out to explain the hazards so that later discussions on the legislative framework can be put into context.

The main hazards associated with the use of electricity are:

■ electric shock;
■ electric burns, both from current passing through the body and from the effects of arcing;
■ the effects of fire that has an electrical origin;
■ the effects of an explosion that has an electrical source of ignition;
■ the effects of electromagnetic radiation.

This chapter is not concerned with other types of injuries, such as crushing and shearing injuries, that may occur as a consequence of, for example, a machine operating aberrantly because of a fault in its electrical control system. However, the safety integrity of electrotechnical machinery control systems is considered in Chapter 13.

Although each of the hazards is considered separately, in reality they often occur together. For example, current passing through the body causes an electric shock, but current passing through tissues also causes the tissue to heat up, often to the point at which burn injuries are caused. So shock and burn injuries are frequently, but by no means always, experienced together.

ELECTRIC SHOCK

If electric current of sufficient magnitude and duration passes through the body it may disrupt the nervous system, causing the painful sensation of electric shock. The effects of such a shock can range from being mildly unpleasant to being fatal, with death most commonly resulting from cardiac arrest or ventricular fibrillation of the heart. Incidents are most frequently caused by the injured person making simultaneous contact with a live conductor and a conductor that is earthed, although some incidents occur when contact is made with two conductors energised at different potentials, neither of which is at earth potential.

Shock injuries are almost always associated with alternating current, with direct current injuries being rare. This is partly because a.c. systems dominate in the workplace and in the home, but also because the excitation effects of direct current on the nervous system are less severe than those of the equivalent r.m.s. magnitude of alternating current. For this reason, only the effects of current at the most common frequencies of 50 Hz and 60 Hz will be considered in detail.

The seriousness of the physiological effects of current passing through the body is directly related to the current's magnitude and duration and to the path that the current takes through the body. The magnitude of the current is related to the voltage across the body and to the body's impedance. Impedance, rather than pure resistance, must be used because the body contains capacitive reactance at power frequencies.

There has been some research carried out over the years to characterise the effects of current flowing through the body. Most of this has been carried out on animals and human cadavers, although some has been carried out on a very limited number of human volunteers. The small number of volunteers is, perhaps, quite understandable. Some general conclusions can be drawn from the published information, much of which is very usefully summarised in British Standard PD 6519 Parts 1 and 2 Guide to effects of current on human beings and livestock; this standard is itself derived from IEC Standard 479.

Body impedance

At power frequencies (50 Hz and 60 Hz) the body impedance comprises resistance and capacitive reactance. The capacitance is concentrated in the skin, whilst the resistance is associated with the skin and the internal path through the body; however, the largest contribution of resistance is the skin barrier. The instantaneous value of impedance for any individual depends on a wide range of factors, not least of which is the applied voltage. In general terms, the higher the voltage the lower will be the total body impedance, with the asymptotic low value being in the order of 750 ohms for 50% of the population for touch voltages greater than 1000 V for a current path from hand-to-foot. The impedance falls as the voltage increases because, at higher voltages, the skin barrier breaks down. At 230 V, which is the voltage at which most electric shock accidents occur, the impedance ranges between 1000 ohms and 2500 ohms for most of the population, again for a hand-to-foot path.

The path that the current takes through the body has a significant effect on the impedance. For example, the impedance for a hand-to-chest path will be in the order of 50% of the impedance for a hand-to-foot path. If the impedance is low, the current will be equivalently high for the same applied voltage.

Other factors affect the impedance. For example, if a sharp object such as a strand of wire punctures the skin, the impedance will be lowered. It will also be lowered by increasing the area of contact, and by wetting the skin surface. Some of the experiments with volunteers used cylindrical metal electrodes, about 100 mm long by 80 mm diameter, held in each hand. It was found that wetting the hands with tap water lowered the impedance a little on the value measured when the hands were dry, but wetting with a 3% salt solution halved the impedance value. Other factors such as age, sex and state of health also affect it.

Effects

The physiological effects of current passing through the body from hand to hand are summarised in Table 1.1. The figures are a rough yardstick and apply to an average person in good health and for a sustained shock exceeding a duration in the order of 1 s. For a duration less than 1 s, greater currents can be tolerated without such adverse reactions. This explains why residual current devices (RCDs) provided for personal protection are rated

Table 1.1 The effect of passing alternating current (50 Hz) through the body from hand to hand	
Current (mA)	**Physiological effect**
0.5–2	Threshold of perception.
2–10	Painful sensation, increasing with current. Muscular contraction may occur, leading to being 'thrown off'.
10–25	Threshold of 'let go', meaning that gripped electrodes cannot be released. Cramplike muscular contractions. May have difficulty breathing leading to danger of asphyxiation from respiratory muscular contraction.
25–80	Severe muscular contraction, sometimes severe enough to cause bone dislocation and fracture. Increased likelihood of respiratory failure. Increased blood pressure. Increasing likelihood of ventricular fibrillation (uncoordinated contractions of the heart muscles so that it ceases to pump effectively). Possible cardiac arrest.
Over 80	Burns at point of contact and in internal tissues. Death from ventricular fibrillation, cardiac arrest or other consequential injuries.

to trip within 30 ms – they do not restrict the level of current flowing, but they do restrict the duration of the current to a nominally safe level. It should also be noted that the effects depend on the path that the current takes through the body. For example, a shock from hand to feet is more dangerous than a shock from hand to hand. PD 6519 defines a heart-current factor that can be used to determine the different effects for different paths through the body.

The heart is vulnerable to ventricular fibrillation during the first part of the 'T' wave, which is approximately 10 to 20% of the cardiac cycle. The shock stimulus produces extra systoles (heart beats) and, if it embraces the vulnerable period, and particularly if it persists over several cardiac cycles, the risk of ventricular fibrillation is increased. It has been found from experiments with dogs, and allowing for the difference in weight, that humans should be able to tolerate 650 mA for 10 to 80 ms and 500 mA for 80 to 100 ms without much risk of ventricular fibrillation. The safety limits, in terms of current versus duration, are depicted in BS PD 6519, to which the interested reader is referred.

Even if the current is not high enough to cause irreversible damage to the heart, it may be high enough to cause muscular contractions, and this is a common cause of injury. For example, it is not uncommon for somebody working on a ladder who receives an electric shock to be thrown off the ladder by the muscular contraction and to be injured by the effects of falling from height.

Although there is a tendency to concentrate on cardiac effects, current flow can also lead to respiratory paralysis. Immobilisation of the respiratory muscles, with the possibility of asphyxial death, can occur if the current is in the order of 18 to 30 mA for a limb-to-limb path (*Electrical Injuries; Engineering, medical and legal aspects*; Nabours, Fish and Hill; Lawyers & Judges Publishing Company; 2000).

The seriousness of electric shock injuries is related to the magnitude of the shock current, but the voltage also needs to be considered since it is this parameter that most people will be aware of. There is no doubt that sustained shocks at the normal mains voltage of 230 V can be, and frequently are, fatal. It is generally accepted that, in dry conditions, 50 V a.c. is a voltage that for most healthy people will not lead to fatal injuries, although a margin of safety is needed, as reflected in the value of 25 V used for the safety extra-low voltage (SELV) systems described in Chapter 3. The risk of serious injury increases as the voltage increases above 50 V, although it is not a step change at that level. Welders, for example, frequently touch their live welding electrodes which have an open circuit voltage up to about 80 V; whereas they will experience minor shocks leading to muscular contraction, the incidence of serious electric shock injuries at that voltage is very low.

Unusual waveforms

The discussion so far has concentrated on a.c. waveforms at the very common power frequencies of 50 Hz and 60 Hz and at the public supply voltage of 230 V single-phase. This is reasonable because by far the large majority of electrical accidents occur on systems energised at those voltages. However, not all electrical systems operate at power frequencies, so some assessment of the risk of injury needs to be made for them.

A good example is electric fence energisers, which are commonly used to power electric fences for stock control and, increasingly, for security. These energisers typically transmit pulses on to the connected fence wire with peak voltages in the order of 10000 V, with a duration of 1 ms, and a pulse repetition frequency of 1 Hz. The idea is that any animal touching the fence will experience an electric shock of such magnitude that it will be deterred from touching the fence again and will therefore remain in its designated field. Of course, it would be highly undesirable for the animal to be electrocuted and it would also be unacceptable if there were to be an appreciable risk of electrical injury to any humans who may inadvertently touch one of the electrified fence wires. Waveforms such as this cannot be assigned the common safety-related limitation of current and duration. The approach that has been adopted by standard makers is to consider the amount of energy needed to cause fibrillation. Thus, in the case of fence energisers, a safe limit of 5 joules per pulse has been set, together with limitations on pulse width and pulse repetition frequency. This is the amount of energy delivered into a 500 ohms load, where that figure of resistance represents the low-end value of resistance for most people for the high voltages at which fence energisers operate.

Guidance on how to assess the risk from non-sinusoidal waveforms is published in Part 2 of BS PD 6519.

Treatment

When somebody experiences an electric shock, it is important that the person rendering first aid should first remove the cause by switching off or otherwise breaking contact between the victim and the live conductor. Care must be taken to ensure that the rescuer does not make contact with the live parts, including the victim's skin, and thereby become a victim as well.

If the victim is suffering ventricular fibrillation, the only effective way to restore normal heart rhythm is by the use of a defibrillator. In that respect, the increasing availability of such units in some places of work, and in public places such as shopping centres and railway stations, is a welcome development. Unfortunately, in most accident scenarios a defibrillator is not immediately available. The first aider should therefore carry out artificial respiration and cardiac resuscitation until either the victim recovers or professional assistance arrives.

It used to be the case that an electric shock treatment placard had to be exhibited in most industrial premises, but this is no longer a legal requirement. However, it is good practice to exhibit such a placard in substations and in places where live work, such as fault rectification and testing, is being carried out.

CONTACT BURNS

Current flowing through the body can cause burn injuries at the points of contact and deep-seated burns in the muscle tissues. The extent of contact injuries is determined by the current density at the point of contact; the higher the current density the more severe will be the injuries. Nabours *et al.* report that the estimated minimum current necessary for first degree burns is 100 mA for a period between 1 and 9 seconds. They observe that, at these currents and durations, cardiac effects will occur before most people experience significant burns.

At low voltages, including 230 V a.c., it is uncommon to see significant burn injuries, although that is not to say it does not happen, as exemplified by the case of an electrician who, when working live, had 230 V applied between the hand and his elbow on one arm, leading to the loss through burning of all the tissues in his forearm. It is not too surprising that he should have suffered such severe burn injuries; the resistance of his forearm would have been in the order of a few hundred ohms, leading to a current of an amp or so flowing. As he got progressively burned the resistance would have reduced due to the skin barrier breaking down and carbonising, leading to an increase in the current and the severity of the heating.

The most usual indication that current has flown is small white blister-like marks on the skin, which often indicate the entry and exit points for the flow of current. At higher voltages, especially in incidents where somebody has come into contact with an overhead high voltage power line, the burning can be severe and can be the main cause of death. A quite common occurrence is for there to be no significant burn marks at the point of entry of the current, but very significant burning or charring of the skin where the current exits the body. An electrician who inadvertently touched a live 240 V a.c. terminal on a temperature controller with a finger of his left hand was also simultaneously leaning against some earthed copper pipework. He could not let go when the shock current flowed from his finger to his shoulder and he was electrocuted. There were no burn marks on his finger, or any other indication of current flow at that location, but he suffered severe charring of the skin where the current had left his body at the point of contact with the pipe.

ARC BURNS

Arc burns are commonly associated with the failure of insulation in electrical equipment, leading to an arc developing in the air between adjacent conductors. A common cause is metal objects such as screwdrivers and spanners shorting a phase conductor to earth, or shorting across conductors at different voltages. The typical consequence of this is the expulsion from the short circuit of a highly energetic arc and hot gases, with temperatures in the plasma typically exceeding 1000°C. The environment will often also contain vapourised metal.

Anybody standing in the immediate vicinity of the arc will suffer burn injuries, most commonly to the face, upper chest and hands, which are often very severe and life threatening. Experience shows that the burn injuries to the face heal significantly more quickly than the burns to the hands.

The power in the arc is determined by the fault level in the system, frequently quoted in million volt amperes (MVA). This is a measure of the amount of current that can be fed into the fault, which is determined by the voltage and the impedance in the fault circuit. Modern systems, even at low voltage, often have very high fault levels.

FIRE INJURIES

Electrical systems that are poorly designed, or have certain fault conditions, may overheat or generate arcs and sparks, to such an extent that they may ignite adjacent flammable materials. Hot spots in circuits can develop when poor connections have high resistance and when circuits are overloaded. Incidents of this type are not particularly common in workplaces, although in 1995 losses due to serious electrical fires represented some 18% of the cost of all serious fires.

EXPLOSION INJURIES

There are many locations in industry that have flammable or potentially explosive atmospheres. For example, petrochemical sites such as refineries, printing works with highly flammable inks, and motor vehicle paint shops are premises in which there will be zones that may contain flammable atmospheres. Any electrical equipment installed in those zones may act as an ignition source unless precautions are taken. If the equipment were to ignite the atmosphere, causing an explosion or fire, people in the vicinity might suffer burns or other trauma.

ELECTROMAGNETIC RADIATION

Consideration of electromagnetic radiation's health effects spans the spectrum from ultra low frequencies associated with the electric and magnetic fields emanating from overhead power lines, through the radio spectrum, up to the infrared, visible, ultraviolet, X-ray and gamma ray ends of the spectrum. There is much controversy about the potential for ill health from exposure to power frequency radiation (50 Hz) and the radiation from cellular phones. Nobody has yet demonstrated or proved a causal link but, because of uncertainty, a considerable amount of research is being carried out internationally to try to determine whether or not there is a link between these forms of non-ionising radiation and ill health. What is well known and can be said with certainty is that overexposure to infrared, ultraviolet, X-ray, and gamma radiation causes ill health, the form of which is well documented.

In the radio spectrum, covering the band 3 MHz to 30 GHz, electromagnetic fields can cause heating of the body tissue by energy absorption, in much the same way that microwave ovens use radio waves to heat food. The eyes are particularly vulnerable to damage because of their water-like composition and because there is no blood circulation to assist in heat dispersal. The limits for exposure to electromagnetic radiation are set down by the National Radiological Protection Board (NRPB), an agency of the Department of Trade and Industry.

Chapter 2

Electrical Accidents and Dangerous Occurrences

INTRODUCTION

This chapter explains the main types of accidents and dangerous occurrences that arise from electrical systems. It also provides some statistical information on incidents, mainly those that occur during work activity.

Electrical accidents at work, in common with other health and safety incidents, must be reported to the appropriate enforcing authority if the legal reporting criteria are met. These criteria are laid out in the Reporting of Injuries, Diseases and Dangerous Occurrences Regulations 1995, which place duties on employers and the self-employed to report certain types of incidents involving employees, the self-employed and members of the public. The enforcing authorities are the Health and Safety Executive (HSE) and local authorities, depending on the type of premises in which the incident occurs. For example, an electrical accident in a shop or office would be reported to the appropriate local authority, whereas an incident in a factory or hospital would be reported to the HSE. Until late 2001, the only way in which accidents and dangerous occurrences could be reported was by submitting a Form 2508 to the enforcing authority. However, there is a now a variety of means for reporting incidents, including on an internet-based reporting website.

Some forms of electrical incident, such as the explosive failure of an item of high voltage switchgear belonging to an electricity distribution company, have to be reported to the DTI's engineering inspectorate. This reporting duty is covered by the Electricity Supply Regulations, as explained in Chapter 8.

Electric burns, and electric shock combined with burns, account for most of the work-related reported electrical accidents. Electric shocks on their own, although very common occurrences, are infrequently notified to the enforcement authorities. This is because an electric shock injury only needs to be reported if it results in death or unconsciousness, or in the injured person being detained in hospital for 24 hours or longer, or in the person being off work for over three days or longer. So, whereas many people experience the immediate effects of a shock such as pain or being 'thrown off' equipment, in most instances there is rapid recovery and the reporting criteria are not met.

Fires and explosions, other than arcing and flashover, constitute a small minority of the reported incidents. In most cases no one is injured and the incident is classified as a dangerous occurrence if the reporting criteria for such incidents are met; this mainly means that the incident should be reported if the fire or explosion results in the plant being out of use for 24 hours or more. There is a mandatory requirement to report incidents which involve unintentional contact with an overhead power line, but contact with an underground cable does not need to be reported unless it results in a fire or explosion and loss of supply for 24 hours or more.

ELECTRIC SHOCK ACCIDENTS

Accidents involving an electric shock are usually subdivided into two categories – direct contact and indirect contact shocks. The standards that will be considered later use this distinction. A direct contact shock occurs when conductors that are meant to be live, such as bare wires or terminals, are touched. An indirect contact shock is usually associated with touching an exposed conductive part that has become live under fault conditions; an example of an exposed conductive part would be the metal casing of a washing machine.

The majority of direct and indirect contact electric shock and burn accidents occur at 230 V on distribution systems or on connected equipment. There are many instances in which high voltage overhead lines are touched, so this is a form of direct contact; however, they usually result in predominantly burn injuries rather than electric shock.

Figure 2.1 illustrates, on the right, the principles of direct electric shocks. The figure shows a typical high voltage to low voltage distribution transformer feeding a number of circuits. Note that the star point of the windings on the secondary side of the transformer is connected to earth, which is the case with all distribution transformers supplying customers of electricity supply companies in the UK. This creates an earth-referenced system, with the neutral of the system connected to this earthed star point.

In Fig. 2.1 outline figure A is experiencing a hand-to-hand shock between a 230 V phase conductor and the neutral of the supply – this could be on, for example, a distribution circuit, or in a distribution board, or in equipment fed at 230 V single phase. Outline B is experiencing a hand-to-hand shock at 400 V between two phases of a three-phase supply; this type of incident is not

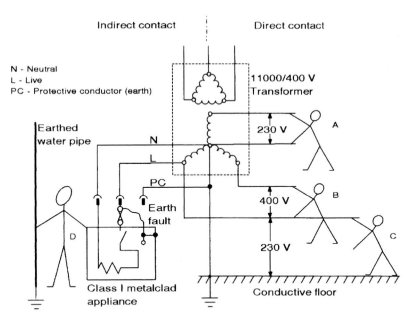

Fig. 2.1 Direct and indirect contact electric shock.

very common. Outline C is experiencing a hand-to-foot 230 V shock between one phase of the supply and earth through a conducting floor (such as a concrete floor); recall from Chapter 1 that a hand-to-foot shock has more severe effects on the heart than a hand-to-hand shock of the same magnitude and duration.

Experience shows that Outline C is experiencing the most common form of direct contact injury: a phase to earth electric shock. A typical incident would be a person picking up a cable on which the sheath and insulation is damaged and touching the exposed live conductor; a shock would be experienced between the hand and feet. Another type of incident would be an electrician working inside a live electrical panel, making inadvertent contact with one of the live terminals and experiencing a shock between the hand touching the live terminal and any other parts of his body in contact with earth.

Outline figure D, on the left of Fig. 2.1, is experiencing a typical indirect contact electric shock. The diagram illustrates the protective (earth) conductor in the flexible cord supplying a metal encased appliance such as a washing machine, coming adrift from its terminal in the plug and touching the phase (live) terminal. The safety issues associated with the wiring of a plug are illustrated in Figs 2.2 to 2.7, although it should be noted that the conductor end soldering shown in Fig. 2.7 is not recommended for use with

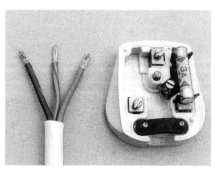

Fig. 2.2 Flexible cord conductors as usually prepared for connecting a 13 A plug.

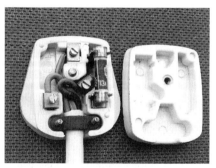

Fig. 2.3 As usually connected. Note the slack in the phase and neutral conductors.

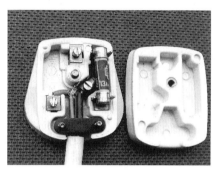

Fig. 2.4 The cord sheath pulled out of the cord grip. The protective (earth) conductor pulled out of its terminal and some of its wires touching the phase terminal.

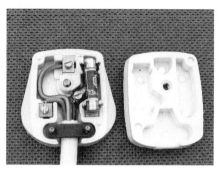

Fig. 2.5 A better way of connecting the plug by adjusting the length of the conductors so that if the sheath is pulled out of the cord grip, the phase and neutral conductors will be disconnected before the protective conductor.

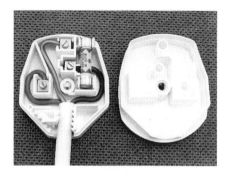

Fig. 2.6 A better plug design. When the plug top is fitted, the internal partitions overlap so that a loose protective conductor cannot touch any 'live' parts.

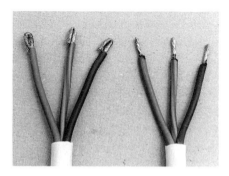

Fig. 2.7 Flexible conductors prepared by soldering or bending over to improve their mechanical strength and conductivity when secured in the terminals.

pinch screw or other non-sprung terminals because the cold flow properties
of solder may lead to a loose connection.

Under the fault condition shown, the casing of the washing machine will
no longer be earthed and it will become live at the full mains potential of
about 230 V. When the person depicted simultaneously touches the live
casing of the heater and an earthed object, such as the water pipe shown, a
hand-to-hand mains voltage shock is suffered. The flow of current in this case
of the indirect contact scenario is depicted in Fig. 2.8 and the equivalent
circuit is shown in Fig. 2.9.

It may seem at first sight that this type of indirect contact scenario is rather
unlikely to occur. However, it is in fact a surprisingly common occurrence.
What tends to happen is that a break in the earthing circuit will occur first,
maybe as a result of the earth connection at a terminal becoming loose, or the
earth connection not being made correctly in the first place, or an open
circuit fault occurring in the protective conductor. This type of open circuit
earth fault will not be detected unless the circuit is tested and the fault

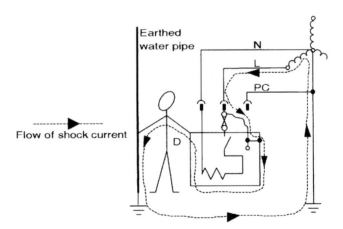

Fig. 2.8 Current flow in case of indirect contact.

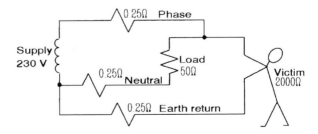

Fig. 2.9 Equivalent circuit of indirect contact shock.

revealed; the fault can therefore exist for some considerable time before it is found. If, in the intervening interval, a phase-to-earth fault occurs as a result of an insulation failure, the conditions for an indirect contact electric shock are created.

It should be noted that in this type of situation the earth fault will probably not be apparent until an electric shock is experienced. This is because the appliance will operate as normal, despite the fact that the exposed metal-work, which should be at earth potential, has become live at mains voltage.

The phase voltage at the substation transformer will be a little higher than 230 V to allow for the inevitable voltage drop in the distribution cables. In urban areas, the line/neutral and the line/earth loop impedances will be comparable and will probably be only a small fraction of an ohm, whereas the victim's hand-to-hand impedance will be in the order of 2000 ohms. Under these circumstances the effects of the circuit impedances can be ignored. The victim's touch voltage will be about 230 V and, for a total body impedance of 2000 ohms, the shock current would be $230/2000 = 0.11$ A. This is high enough to cause ventricular fibrillation in many people should the current flow for about 0.5 s.

This type of accident, which requires two faults to occur (the loss of the earth connection followed by a phase-to-earth fault), is common. It is interesting to observe that our electrical systems are designed in such a way that two faults such as this can lead to injury and death. It is for this reason that preventive maintenance of electrical systems is very important – if we were not to maintain our systems, the rate of occurrence of these two faults and the consequential injuries would be far higher.

The nature of the fault described means that the touch voltage is in the order of 230 V. However, many indirect contact shock accidents occur at less than mains voltage. This can be quite fortunate for the injured person because the shock current will be lower, thereby reducing the adverse effects and improving their chances of being able to let go of the conductors and survive the incident.

In order to explain how the voltage is reduced, consider the fault shown in Fig. 2.10 where there is a short circuit between the centre of, say, a heating element and the appliance's metal case. If the appliance has lost its earth connection because, for example, the earth core in the flexible cable has become detached from its plug terminal, the metal case will acquire the potential to earth of half the supply voltage, about 115 V. This will be the shock voltage for the person who is in contact with the case and the earthed water pipe. At this voltage the victim's body resistance will be greater than it would have been at 230 V, so the shock current would be less than half that corresponding to a touch voltage of 230 V.

If the appliance has not lost its connection to earth and the water pipe is

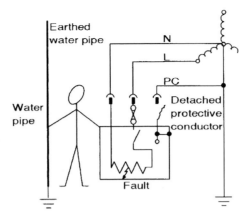

Fig. 2.10 Shock at voltage lower than mains voltage.

properly bonded to the installation's main earth terminal to form an equipotential zone (as described in Chapters 3 and 10), the potential difference between the appliance's metalwork and the water pipe would be very small and the person would not experience an electric shock. The equivalent circuits associated with these two situations are shown in Figs 2.11 and 2.12.

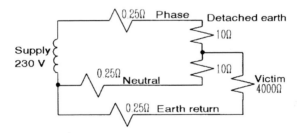

Fig. 2.11 Equivalent circuit of Fig. 2.10.

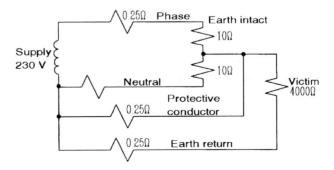

Fig. 2.12 Equivalent circuit of Fig. 2.10 with the protective conductor intact.

Portable tools

The discussion so far has concentrated on incidents associated with fixed or transportable equipment and appliances. Whereas these represent a significant proportion of incidents, many electric shock accidents arise during the use of portable and hand-held power tools. These accidents are commonly linked to a failure of the insulation, mostly on the cables supplying the tools. This is especially prevalent in harsh environments, such as construction sites and farms, where the risk of mechanical damage to the sheathing and basic insulation is especially high. Insulation damage will often expose the live conductor in the cable. When this is touched, a hand-to-foot shock will be experienced. If the live conductor is being gripped at the instant the shock current starts to flow, the victim may well not be able to let go.

Since portable power tools are frequently used on ladders, scaffold structures and platforms, a relatively common result of a person experiencing an electric shock when using a power tool under these working conditions is a fall from height leading to impact injuries.

Extraneous metal parts

Extraneous metal parts are conductive parts that are not part of an electrical installation, such as metal gas pipes, reinforcing bars and window frames. Sometimes they can become live and cause electric shock accidents. An example of a reported case involved a metal ladder on a construction site. The ladder was lashed to a wooden scaffold board near its top and had its feet resting on another wooden board. Trapped under one foot was a damaged PVC mains cable, with a phase conductor touching the foot, thereby making the ladder live at 230 V. A labourer, wearing insulating rubber boots climbed the ladder and, when near the top, grasped a metal scaffold pool with one hand while holding on to the ladder with his other hand. The bottom of the scaffold pole was in a muddy puddle which provided an effective earth for the return current. The victim was subjected to a 230 V phase-to-earth hand-to-hand shock and was electrocuted.

Although the scaffold pole was a poor earth electrode, having a considerable earth resistance, it allowed sufficient current to flow through the labourer to cause fatal electrical injuries. Typical resistances would be 400 ohms for the earth connection and 2000 ohms for the labourer's body, resulting in a shock current of 230/(400 + 2000) = 0.096 A. A hand-to-hand current of 96 mA is sufficient to cause ventricular fibrillation.

BURN ACCIDENTS

Electrical contact burns

Many electric shock incidents such as those described are accompanied by minor burn injuries at the points of contact. These are frequently characterised by small blisters at the entry and exit points. More severe burn injuries arise when large currents flow through the body, such as when a high voltage conductor is touched.

Arc burns

Arc burn injuries, also known as flashover injuries, are mostly sustained when work is being carried out on live electrical equipment, most commonly during fault finding and testing activity or when equipment is being added to existing live equipment. For example, adding a new circuit into an existing distribution board may involve making connections between the busbars, fuse carriers or circuit breakers, and neutral links. There is only limited clearance between the live parts on different phases and the earthed metalwork of the distribution panel. Short circuits between phases, or between a phase conductor and earthed metal, occur when a metal component accidentally bridges the phase conductors or bridges a phase conductor to the earthed metal frame or casing.

As explained in Chapter 1, the effect of the short circuit is to cause a very large current to flow, the magnitude of which depends on the circuit voltage and impedances but which may be in the order of thousands of amps, especially if the short circuit occurs close to the distribution transformer. The heating effect of the current vapourises the metal. The gaseous products create a conducting path between the conductors, supporting the flow of current between the conductors. The air breaks down, creating an ionised plasma. The net effect is an explosion of extremely hot gaseous products and flames, accompanied by a blast wave, that evolves from the point of the short circuit, enveloping anybody in the near vicinity. The explosion will last until either the circuit protective device operates or the current is extinguished by the destruction of the circuit.

The typical injuries are burns to the face, neck and chest. The backs of the hands are also frequently badly burned as they are quickly raised to protect the face. In some cases the burns are sufficiently deep and extensive to be fatal, although in most cases a recovery is made.

The short circuit can be caused by any conducting material. For example, in one recent case a number of electricians were installing a new cable on a cable tray adjacent to a 400 V distribution board. They were using single strands of wire to tie the cable to the tray, with the wire lengths typically in the order of 0.75 m. As one of the electricians was manipulating one of the pieces of wire, unbeknown to him one end of it entered a small gap in the cover of the live distribution panel. It created a short circuit between the earthed cover of the panel and a 230 V terminal on an air circuit breaker; the fault level at that point was 15 MVA. The ensuing explosion blew the cover off the panel and the flames severely burned the electrician on his back.

Short circuits have also been caused by uninsulated tools which, when carelessly used, have bridged between live parts or between a live part and earthed metal. Another common cause is the use of inadequately insulated instrument probes. The bare section of the metal probe, if carelessly used, may bridge between a live part and earthed metalwork or between live parts of different polarity, again causing a short circuit with the consequential arcing and burn injuries.

Electricians taking short cuts by unnecessarily carrying out live work is another cause of burn accidents. An accident of this type occurred on a 100 A triple pole switchfuse (see Fig. 2.13) which was part of a low voltage (400 V) switchboard and which was used as an isolator for a furnace control panel. The control panel developed a fault and two electricians were called to repair it. One of the electricians went to the control panel and the other to the

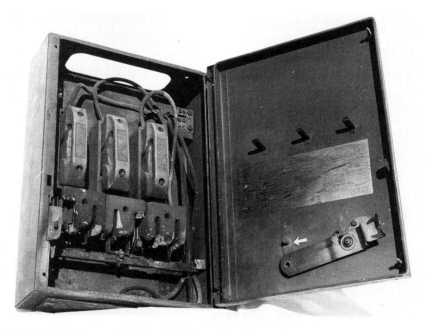

Fig. 2.13 The damaged 100 A switchfuse. The remains of the moving contact are still engaged with the fixed contacts. The arrow indicates the hinge pin. The operating lever is below it.

nearby LV switchboard where he operated the handle of the switchfuse to the 'off' position and the switch did not open. He opened the door and found that the operating lever had become disengaged from its pivot pin and the moving contacts were still engaged with the fixed contacts. A little thought would have indicated that to isolate the circuit he could have either removed the fuses or he could have gone to a main switchboard, some 40 m away, and opened a 600 A switchfuse. He chose, however, to use an 11 in screwdriver as a makeshift operating lever to open the switch. The screwdriver slipped and the blade touched the nearby live moving contact, causing a phase-to-earth fault which spread to involve all three phases (see Figs 2.14 and 2.15). The fault current was about 4000 A and the flame that was emitted caused severe burn injuries to the electrician.

Fig. 2.14 The slipped screwdriver in the approximate position where it caused the arcing.

Fig. 2.15 The arc damage to the screwdriver blade.

FIRES

Wiring faults

The experience of many forensic investigators is that faulty wiring is often blamed for fires in buildings where the cause of the fire is unknown. This may be because fire prevention officers have to state the probable cause of the fire in their report form. If the fire has destroyed the evidence, which is not unusual, there is a temptation to attribute the fire to a cigarette end or faulty wiring.

A fire which occurred in the early hours of the morning in a small wooden factory building was attributed to faulty wiring, probably because the owner admitted his wiring was old and needed replacing. The building was completely destroyed by the fire so there was no evidence of the cause left. An investigation, however, revealed that the owner took the precaution of isolating his installation when he left at the end of the day, by opening the main switch which was situated by the entrance. The only wiring that was energised during the night was the meter tails. The supply company's armoured service cable, the meter and main switch survived the fire, the switch was found to be open and there were no signs of arcing on the meter tail conductors. It was, therefore, improbable that the wiring had been the cause of the fire and the attribution was not justified.

HSE statistics attribute very few fires in factories to faulty wiring. It is probable, therefore, that faulty wiring causes some fires but is not a major cause.

Lead-sheathed vulcanised-rubber-insulated cable

Wiring faults which can cause fires are generally due to defective insulation or bad connections. Before World War II, a favoured method of house wiring used lead-sheathed, twin vulcanised-rubber-insulated (VRI) conductors which were concealed beneath the wall plaster, and run below the floorboards and in the roof space. For earth continuity at junctions, bridging conductors and clamps on the lead sheath were used. In the course of time, these clamps tended to slacken and the joint to corrode, which increased the impedance of the protective circuit. The natural rubber insulation perished. The rubber hardened and cracked and separated from the conductor if disturbed. Under these conditions, dampness or conducting foreign matter would cause leakage between the phase conductor, earthed sheath or neutral, or the phase conductor, if disturbed, might contact the lead sheath. The leakage currents might be insufficient to cause immediate fuse failure but might well cause sparking at the bad contact points which could, in turn, ignite anything inflammable in the vicinity and cause a fire.

Most of this wiring has been replaced but there is still a fair amount installed in older buildings.

Tough-rubber-sheathed and PVC cable

Lead-covered wiring was superseded by the tough-rubber-sheathed (TRS) system where the VRI conductors were contained in a tough rubber sheath with an uninsulated protective conductor. This solved the earth continuity problem but the wiring became a fire hazard when the rubber perished. After World War II, PVC was introduced and replaced the rubber for both insulation and sheath and, apart from some teething troubles in its early years, it is now satisfactory and appears to have a very long life. As it burns less readily than rubber it is a lesser fire hazard.

The three foregoing types of wiring are all vulnerable to mechanical damage which may cause short circuits or leakage and lead to ignitions. PVC-insulated cables in metal conduit or trunking are better protected and are much less of a fire risk because in the event of a fault the metal enclosure is a fire barrier.

Mineral-insulated, metal-sheathed cable

Mineral-insulated, metal-sheathed (MIMS) cable is suitable for use in most environments, but it is not so suitable for use where there may be vibration. In an old paint spray booth which predated the Highly Flammable Liquids and Liquefied Petroleum Gases Regulations, a spider-frame-mounted, electrical extractor fan had been positioned at the mouth of the extract duct and connected by MIMS cable. The fan motor was totally enclosed and explosion-protected as the paints used were flammable. The maintenance was poor and paint accumulations were allowed to build up on the fan blades causing an imbalance and vibration. This led to embrittlement of the copper sheath of the cable, which cracked and broke. Then one of the conductors fractured and sparking occurred between the ends. Ignition of the spray started a fire in the paint accumulations on nearby surfaces and the factory was burnt down.

Bad connections

Bad connections in terminals, due to loose screws and poor contacts in socket receptacles or fuse carriers, can lead to local sparking which, in turn, causes corrosion, localised overheating of conductors and possible ignition of anything flammable nearby.

Overloading

> Overloading of conductors is not a frequent fire causation because the excess current protection is usually matched to the conductor size and will operate before the conductors are seriously overheated.

Overheating

> Localised overheating can occur in badly designed tungsten lamp luminaires from the radiated and conducted heat from the filament and in hot environments where the cables have not been derated or suitably insulated for the high ambient temperature. The increase in the use of thermal insulation, to reduce heat losses in buildings, can be a problem as cables covered by thermal insulation have to be derated to avoid overheating. Any overheating of conductors damages most insulating materials and enhances the fire risk.
>
> The winding insulation in electric motors of the enclosed ventilated type will overheat and may break down if the ventilation is impaired, e.g. by foreign matter clogging the air vents and ducts. The resultant short circuit between turns or to earthed metal may cause an ignition, but may not cause a sufficient increase in the supply current to operate the excess current protection.

Convector heater fire

> In one fire incident, a convector heater had a badly designed resistance wire element. Some of the coils touched, resulting in part of the element being short circuited. The increased current in the rest caused overheating of the wire and a progressive break-up of the element. Some of the hot pieces of wire fell through the slots in the base plate on to the carpet below and set the house on fire. Although the increased current was sufficient to burn out the element, it was not enough to rupture the HBC fuse in the plug top.

Ceiling luminaire fire

> To reduce the heat losses in a bedroom with a flat roof and suspended ceiling, cellulose fibre (torn paper) insulation was blown into the roof space and settled unevenly on the fibre board liner above the plaster board ceiling. Some of it fell through the holes in the fibre board provided for the luminaires, and lodged against the lamp glasses (see Figs 2.16, 2.17 and 2.18). When the lights were switched on, the lamps got hot and caused the insulating material to smoulder. This insulation is chemically treated to prevent combustion. It will smoulder but should go out when the source of

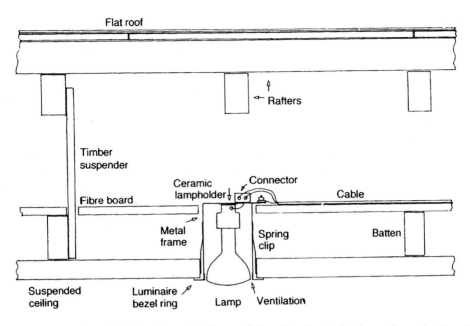

Fig. 2.16 Ceiling luminaire fire. Note: the blown cellulose fibre insulation is not shown for the sake of clarity. Some of it fell through the hole in the fibre board on to the lamp.

ignition is removed, i.e. in this case when the lamps are switched off. The smouldering insulation, however, ignited the fibre board and set fire to the wooden roof structure.

EXPLOSIONS

There is sometimes difficulty in differentiating reported explosions from flashover-type incidents, because the latter can easily be interpreted as explosions. However, reported explosions include events such as underground cables being struck and generating explosive failures, the ignition of flammable atmospheres by arcs and sparks in electrical equipment, and explosions in oil-filled control and switchgear, such as high voltage ring main units and oil-filled transformers.

Oil circuit breaker explosion

In a steelworks example, an 11 kV oil circuit breaker controlled an arc furnace transformer. During the melting process, prolonged short circuits occasionally occurred between the scrap and the furnace electrodes, causing the time-delayed overload protection to trip the circuit breaker. The resultant

Fig. 2.17 The ceiling luminaire with 60 W projector lamp. The clear glass neck of the lamp gets hotter than the curved section below it which is internally silvered.

arcing at the circuit breaker contacts when the fault current was interrupted carbonised the oil. The carbon particles were deposited on the surfaces of the fixed contact insulators and on the earthed metal surfaces of the oil tank (see Fig. 2.19). The deposit formed a conducting path and eventually a flashover occurred between phases and between a phase and the earthed metal tank. The high energy arcing rapidly vaporised the oil and the gas pressure burst open the tank, scattering burning oil. The force of the explosion tore the switchroom's steel door off its hinges and blew it 6 m across a gangway.

In similar events elsewhere, the force of the explosion has been known to demolish whole substations. Indeed, in a recent incident in a gasworks, the explosion in the current transformer chamber of an 11 kV switchboard occurred while engineers were inside the substation. Fortunately, they heard the switchgear 'fizzing' moments before the explosion and had the presence of mind to depart the substation in something of a hurry. The blastwave hit

Fig. 2.18 A mock-up showing the luminaire fitted in the ceiling. The blown insulation fell through the hole in the fibre board on to the glass lamp.

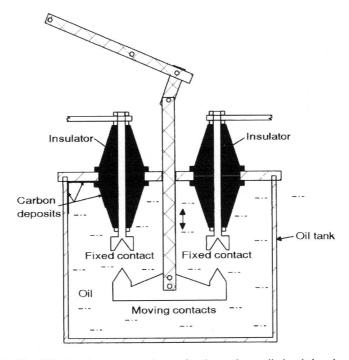

Fig. 2.19 Simplified section on one phase of a three-phase oil circuit breaker.

them just as they were running through the door, resulting in minor burns to the backs of their necks.

Historically, much of the earlier oil-filled switchgear had what was known as dependent manual operating mechanism. In this type of mechanism, the contact closing mechanism was directly linked to the operating handle, such that the closing speed and pressure of the contacts was directly related to the effort being exerted by the engineer operating the switchgear. This should be compared with modern switchgear in which the closing mechanisms are spring-assisted.

If a switch, for example, has dependent manual operation and is being closed on to a fault, and the closure is hesitant, the fault current will cause arcing at the contacts. In some designs, the electromagnetic forces will force the contacts apart, making the arcing worse. Alternatively, the operator may sense the increasing pressure and instinctively reverse the closing operation, again making the arcing worse. The rapid oil vapourisation and consequential internal pressure rise burst the tank and usually envelop the operator with burning oil.

Although the number of oil-filled dependent manual switches and circuit breakers is decreasing, there are many still in service. The HSE has been concerned for some time that today's engineers may not understand the particular hazards associated with using this type of switchgear on modern high voltage distribution systems, and the need to operate it quickly and positively, and the risks associated with reversing the operation when closing on to a fault. Another concern is that some of the very old circuit breakers that are still found in industrial premises have plain break contacts, without the aids of turbulators and other techniques used to assist in the extinguishing of the arc that is drawn when the contacts open. There is also concern that some of the older switchgear may be underrated, and examples of this are still being found. This means that the switchgear's fault breaking capacity, which may have been adequate when it was first designed, may be lower than the very high fault levels that exist on today's distribution networks. This creates the potential for the circuit breaker to fail explosively when clearing a short circuit fault beyond its design capacity. Finally, there is concern that the switchgear may not be properly maintained.

In an attempt to raise awareness of this problem, the HSE produced an Information Document in 1995 to describe both the problem and the action that needed to be taken to reduce the risks to an acceptable level.

Although dependent manual and underrated circuit breakers are a significant hazard, they are not the only cause of accidents involving oil-filled switchgear. In another tragic accident, in 1997, two electrical engineers were killed when they were working on an oil-filled 11 kV ring main unit (RMU) that was connected into a power company's high voltage distribution network.

An RMU is a device containing two switches and a tee-off fed through a fused switch or a circuit breaker; the tee-off circuit normally connects on to a transformer which steps the 11 kV supply down to 415 V for local power distribution. In oil-filled RMUs, of which there are many thousands in use in the UK, the switching and power distribution components are immersed in oil. A schematic circuit diagram of a typical RMU, omitting various earth connections, is shown in Figure 2.20, and a typical 11 kV external substation incorporating an oil-filled RMU (a Long & Crawford T3GF3 unit) is shown in Fig. 2.21. Note that distribution substations are now commonly being installed as packaged units inside weather-proof enclosures with pressure-relief panels, an example of which is shown in Fig. 2.22.

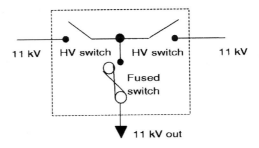

Fig. 2.20 Basic schematic of high voltage ring main unit.

The engineers had been using a set of test probes to connect on to the cable contacts of one of the oil switches in the unit. This is a commonly used means of testing the condition of one of the cables connected into the unit, usually as part of a fault finding routine or to check the insulation characteristics of the cable. It can be done with the other oil switch closed and the tee-off section still live to ensure continuity of supply to the local consumers. As the engineers were withdrawing the test probes from the RMU, a steel guide pin became detached from the test probe assembly and fell into the oil tank. It dropped through the oil to the base of the tank, where it shorted one of the 11 kV busbars to earth. The arc that developed caused a gas bubble to be formed in the oil and the increasing pressure forced the oil out of the tank. The mist of oil was ignited, possibly by the arc, and the engineers suffered fatal burn injuries. Despite the fact that the fault was cleared in less than 0.5 s, nearly 90 MJ of energy had been fed into the oil tank.

Flammable atmosphere explosions

Much emphasis is placed on ensuring that electrical equipment in flammable atmospheres is so constructed that it cannot act as a source of ignition, and

Fig. 2.21 Typical external 11 kV substation incorporating oil-filled ring main unit.

the details of this are explained in Chapter 15. Notwithstanding the use of explosion protected equipment in many places, incidents do occur. For example, the highly flammable paints being used to paint the bilge area of a ship in dry dock were ignited when a 500 W 110 V halogen lamp was introduced into the area to provide illumination for the painters. Unfortunately, the surface temperature of the halogen bulb exceeded the autoignition temperature of the vapours, which ignited and created a fireball. The painters were lucky to escape without serious injury.

Another mechanism for flammable atmospheres to be ignited is by electrostatic discharge. It is not uncommon for highly flammable liquids, such as Toluene, to be dispensed into metal drums as part of a batch production cycle in, for example, the chemical and printing industries. If suitable precautions are not taken, it is possible for significant electrostatic potentials to

Fig. 2.22 Typical modern packaged substation – the switchgear is installed inside prefabricated weather-proof enclosures.

develop on the surface of both the drum and of the liquid. In these circumstances, if an earthed object is brought up to the drum, the static will discharge to earth, creating a spark which in many cases will have enough energy to ignite the vapours and to cause an explosion. This is an all too frequent occurrence that demonstrates the importance of taking precautions against the build up of static charge in flammable atmospheres.

ACCIDENT STATISTICS

It is always helpful to review accident statistics in order to gain an appreciation of the underlying factors that cause accidents. In some ways, the use of the word 'accident' to describe incidents is unfortunate because it tends to hide the fact that many of the incidents in which people are injured are entirely preventable, although it is accepted that most are unexpected and unintentional. A very good source of information on electrical accident statistics is the HSE's publication *Electrical Incidents in Great Britain – Statistical Summary*, published in 1997. Although this covers the period 1989 to 1996, the underlying statistics remain valid today.

During the seven-year period 1989 to 1996, 148 employed and self-employed people were killed at work in the UK from the direct consequences of electric shock and burns. This means that, on average, some 21 people per year suffered fatal electrical injuries. This figure does not include those people who suffered non-electrical injuries that resulted from, for example, falls that were initiated by an electric shock or the effects of an electrically-ignited explosion. To some extent, therefore, the headline figure of 21 fatalities per year does not tell the full story and has to be treated with caution. In addition to these fatal accidents, 6361 people were reported as having suffered non-fatal electrical injuries.

It can be assumed that the numbers for fatal accidents will be accurate because the HSE learns about all workplace deaths, but those for non-fatal injuries will seriously underestimate the actual number of workplace injuries because of endemic underreporting. Many employers and self-employed people are reluctant to report injuries to the HSE, despite the fact that not to do so is a breach of the Reporting of Injuries and Dangerous Occurrences Regulations 1995. Another influencing factor is that the Regulations only require to be reported those electric shock injuries that lead to unconsciousness or require resuscitation or admittance to hospital for more than 24 hours; very many electric shock incidents do not fall within those categories. Otherwise, electrical injuries must be reported if they lead to:

- death;
- any fracture, other than to the fingers, thumbs or toes;
- any amputation;
- dislocation of the shoulder, hip, knee or spine;
- loss of sight (whether temporary or permanent);
- injuries that result in the person being away from work or unable to do the full range of their normal duties for more than three days.

Historically, some 50 to 60% of fatal electrical accidents are due to persons coming into contact with overhead power lines. The most common type of overhead line incident involves construction and agriculture vehicles, such as tipper trucks and excavators, making inadvertent contact with 11 kV and 33 kV overhead lines.

In a recent incident, for example, a tree harvesting vehicle made contact with a 33 kV overhead line and stayed in contact with it because the fault current that flowed to earth through the vehicle caused its tyres to blow out. The fault current was large enough to cause the circuit breaker protecting the line to trip, removing power from the line. The driver of the vehicle had not been injured at all when the vehicle struck the line because, while in his cab, he had effectively been inside an equipotential cage. However, he got out of the vehicle to inspect its engine for damage. As he was inspecting the vehicle, the control engineer of the high voltage distribution network remotely closed the circuit breaker to reenergise the power line. Although the circuit breaker tripped again because the line was still earthed through the vehicle, the vehicle became live for the few tens of milliseconds that it took the circuit breaker to reopen. Current flowed through the driver as he was touching the vehicle at the instant that the line was reenergised and he died of burn injuries. The moral of the story is that, in incidents such as this, the driver of a vehicle that is in contact with an overhead power line should jump clear of the vehicle and stay away from it until engineers from the power company have declared the situation to be safe.

To emphasise the point about the danger from overhead lines, during the ten-year period 1986 to 1996 some 130 members of the public were killed by making contact with overhead power lines.

Fatalities from electrical incidents account for about 6% of all work-related fatal accidents, with the overall accident statistics being dominated by injuries sustained from being struck by vehicles and by slips, trips and falls. In contrast, non-fatal electrical injuries account for just 0.5% of all non-fatal injuries. This is probably because many electric shock incidents go unreported – the consequences of an electric shock tend to be either very severe or the victim is able to walk away from it.

When looking at the type of work being carried out when electrical incidents occur, some 44% of the total happen during maintenance activity. This should not be particularly surprising. Fault finding work, which is a subset of maintenance, often involves people carrying out live work while under production pressure to reinstate equipment, and without structured planning of the work and without formal risk assessments, and in difficult physical conditions. This is the type of situation in which mistakes and consequential injuries are both likely and foreseeable. The fundamental importance of this is that a very large proportion of electrical accidents result from people making mistakes or, more formally, from failures in systems of work. Allied to this, it is instructive that some 50% of electrical accidents occur because the people involved in the activities were not competent. The significant topic of competence will be considered in detail in later chapters.

When considering dangerous occurrences rather than electrical accidents, there are two types of incident that are reportable to the HSE and are therefore of interest:

■ Electrical short circuit or overload attended by fire or explosion which results in the stoppage of the plant involved for more than 24 hours or which has the potential to cause the death of any person.
■ Any unintentional incident in which plant or equipment either:
   □ comes into contact with an uninsulated overhead electric line in which the voltage exceeds 200 volts; or

☐ causes an electrical discharge from such an electric line by coming into close proximity to it.

During the period 1989 to 1996, there were 952 of the former type of incident, representing just under 4% of the total, and there were 1393 of the latter type of incident, representing nearly 5.5% of the total.

Chapter 3
Basic Safety Precautions

INTRODUCTION

The hazards and risks associated with electricity were described in Chapter 1, and the types of electrical accidents that occur were covered in Chapter 2. The aim of this chapter is to introduce the principles of the most common precautions that are available to prevent accidents. It is useful to have an appreciation of the techniques before the legal requirements for electrical safety are addressed, although the techniques will be described in more detail in later chapters in relation to particular applications and to the relevant standards.

PRECAUTIONS AGAINST ELECTRIC SHOCK AND CONTACT BURN INJURIES

The precautions against electric shock described in the following paragraphs also provide protection against the burn injuries that often occur when contact is made with live conductors. The measures can conveniently be grouped into a small number of categories:

- Techniques that aim to prevent live conductors being touched, including:
  - the use of enclosures and insulation;
  - placing conductors out of reach.
- Techniques that aim to limit the amount of current that can flow, and/or its duration, when a conductor is touched, including:
  - the use of reduced and extra low voltages;
  - the use of electrically separated and unreferenced systems;
  - the use of techniques that limit energy and current to nominally safe levels;
  - the use of residual current devices as supplementary protection.

■ Techniques that ensure that electrical systems become disconnected from their sources of energy in the event of faults occurring that may lead to danger.

These techniques are summarised in the following text.

Enclosures and insulation

Direct contact with live parts is most commonly prevented by covering the conductors with suitably-rated insulating material or by placing them inside an enclosure. In the latter case, unless the live conductors inside the enclosure are insulated or placed behind barriers to prevent them being touched when the door is open, additional precautions must be taken. Typically, these will require that the enclosure door must only be capable of being opened using a key or a tool. Alternatively, the door must be interlocked in such a way that either the conductors must be made dead before the door can be opened or the act of opening the door reliably makes the conductors dead. Additionally, a warning symbol should be fixed to the door to highlight the danger.

The majority of fixed apparatus is of Class I construction, as defined in BS 2754 : 1999 Construction of electrical equipment for protection against electric shock. This means that it usually has a metal case enclosing the live parts and that the live parts are insulated from each other and from the carcass by basic insulation only. The case, which is earthed by connection to the protective conductor, protects the internal wiring and components from damage and prevents direct contact.

The extent to which the enclosure prevents the ingress of water and dust is defined using an ingress protection (IP) code set out in BS EN 60529 : 1992 Specification for the classification of degrees of protection provided by enclosures. The IP code typically uses two figures. Although a third can be quoted for mechanical impact, it is not usually used in practice. The first figure shows the degree of protection against solid objects and the second figure shows the degree of protection against liquids. The salient features of this code are set out in Table 3.1. So, for example, an enclosure with an IP rating of 55 would generally be regarded as being suitable for external use. A common code is IP2X, which means that the enclosure is fingerproof but has no rating for protection against the ingress of moisture. The terminal holes in domestic power sockets to BS 1363 are IP2X. It is common practice now to make the internal construction of electrical panels meet the IP2X standard so that the enclosure door can be opened without allowing anybody directly to touch live parts.

Nowadays, insulating plastics are commonly used for wiring accessories such as switches, socket outlets and ceiling roses which usually have no

Table 3.1 The IP code system.

First number: protection against solid objects	
0	No protection
1	Protection against solid objects over 50 mm
2	Protection against solid objects over 12 mm
3	Protection against solid objects over 2.5 mm
4	Protection against solid objects over 1 mm
5	Protected against dust – limited ingress (no harmful deposit)
6	Totally protected against dust

Second number: protection against liquids	
0	No protection
1	Protection against vertically falling drops of liquid
2	Protection against direct spray up to 15° from the vertical
3	Protection against direct spray up to 65° from vertical
4	Protection against direct spray from all directions
5	Protection against low pressure jets from all directions
6	Protection against strong jets
7	Protection against the effects of immersion up to 1 m for short periods
8	Protection against long periods of immersion under pressure

accessible conductive parts. Such enclosures provide virtually complete protection against electric shock provided they are kept dry and clean.

Portable apparatus may be of Class I construction, in which case the metalwork is earthed by a protective conductor which is a separate core in the flexible supply cable, or it may be of Class II construction, having no protective conductor terminal as its metalwork is not earthed. The essential safety feature of Class II apparatus is that the basic insulation is supplemented by additional insulation to provide a further safety barrier and danger arises only if both insulating layers fail. There are two types: 'all insulated' where the supplementary insulation is a plastic case, and 'double insulated' where there is a metal case but live parts are separated from it by two layers of insulation. It is not advisable to use enclosed ventilated Class II apparatus in wet environments, because moisture may penetrate and provide a conductive film between the touchable surfaces and internal live conductors.

Safe by position

Uninsulated conductors energised at dangerous voltages can, in principle, be made safe by placing them out of reach. Perhaps the most obvious example of this technique is the ubiquitous power distribution and transmission overhead lines, operating at voltages of 230 V and upwards, that exist throughout the UK. 11 kV overhead lines, for example, must be suspended at heights no lower than 5.2 m above ground level, on the basis that this is presumed to be high enough to prevent inadvertent contact. It has already been observed that 10 to 15 people are killed each year by making contact with these power lines, so it is reasonable to observe that the technique is not entirely successful. However, the cost of implementing engineering measures to prevent these deaths, such as insulating the conductors or replacing them with buried cables, would be prohibitive and would not be reasonably practicable.

A common example of the technique in factory environments is the bare conductors running at height along a wall, which are used to provide power to an overhead travelling crane; the crane has power pick-offs that run along the conductors as the crane moves, to provide power to the drive motors. This technique is satisfactory so long as there is no easy access to the bare conductors from the crane structure or, for example, from ladders placed up against the wall.

Reduced voltage system

Increased safety is attainable by reducing the potential shock voltage. In the UK the most common system operates at 110 V three-phase or single-phase. In the former, the star point of the supply generator or transformer is earthed, and in the latter the centre point of the output winding is earthed. Reduced voltage is mainly used to supply Class I and Class II portable tools on construction sites, but it is coming into increasing use in factories. The distribution system and apparatus are detailed in Chapter 11, but for factory use indoors the wiring is commonly fixed and the apparatus does not have to be weatherproof nor usually as robust as that employed on construction sites.

Safety extra-low voltage (SELV)

For a greater degree of safety, lower voltages may be used. Opinions differ as to what constitutes a safe low voltage, but in normal dry environments it is generally considered that no harmful shock will result from handling live parts where the voltage does not exceed 25 V a.c. or 60 V ripple-free d.c. between conductors or between any conductor and earth. In wet and/or in confined conductive locations where the normal body impedance is likely to be reduced, lower voltage limits are needed to avoid danger.

For both normal and abnormal environments it is essential to preserve the integrity of the extra-low voltage circuit and prevent it from becoming live at a higher voltage from another system. Therefore, in a SELV system, any exposed conductive parts should not be connected to, or be in contact with, the protective conductor of another system, nor with extraneous metal which could be energised by another system. A step-down transformer, to provide the safe low voltage, should be a safety transformer to BS 3535 or BS EN 60742 Isolating transformers and safety isolating transformers. As the safeguard is the extra-low voltage, which results in there being virtually no shock risk, no enclosures are required and only functional insulation is employed. The lack of enclosures, however, increases the risk of burn injuries from handling live parts in the event of arcing or overheating consequent on a short circuit fault.

Above 25 V a.c. and 60 V d.c. up to 50 V a.c. and ripple-free 120 V d.c. there is a shock risk, albeit a relatively small one in normal environments, so some additional precautions are required. Direct contact with live parts should be prevented by insulation, barriers or enclosures. No part of the circuit should be earthed, and exposed conductive parts should neither be connected to a protective conductor nor otherwise earthed. This ensures that two faults would have to occur to create a shock hazard. Again, if exposed conductive parts have to be in contact with extraneous conductive parts, the latter must not be capable of attaining a voltage exceeding that of the SELV circuit. In abnormal environments, reductions in the upper voltage limits, relative to the degree of risk, are necessary for the same level of protection.

Extra-low voltage other than SELV

BS 7671 : 2001 Requirements for electrical installations – The IEE Wiring Regulations 16th Edition, covered in Chapter 10, recognises two other forms of extra-low voltage: protective extra-low voltage (PELV) and functional extra-low voltage (FELV). The former varies from SELV only by having its circuits earthed at one point only. It has few applications. Protection is provided either by barriers or enclosures to at least IP2X or with insulation capable of withstanding 500 V d.c. for 60 s. Where, however, the voltage does not exceed 25 V a.c. or 60 V ripple-free d.c. in a dry location within the equipotential zone, these additional precautions are not required, but otherwise the limits are 6 V a.c. or 15 V ripple-free d.c.

For FELV, additional safety precautions are cited in BS 7671. These include, for example, that apertures in horizontal top surfaces should be IP4X, i.e. to exclude solid objects exceeding 1 mm diameter, and the enclosure should be stable and strong and need a key or tool for its removal. The insulation should be improved so that it will withstand the same test voltage as applies to the primary circuit. The exposed conductive parts should be connected to the protective conductor of the primary circuit.

Limitation of energy

There are a number of applications where limitation of energy is used as the method of protection. As an example, a common application in the agricultural sector for stock control, and increasingly in the security sector for guarding applications, is electric fences. As explained in Chapter 2, these consist of uninsulated wires, often running to lengths measured in kilometres in the case of stock control fences, which are connected to one or more energiser units. The energiser transmits pulses on to the fence wires, with the peak voltage of each pulse being in the order of 5 to 10 kV, with a pulse duration in the order of 1 ms and pulse repetition frequencies of about 1 Hz. The design aim is to provide a sufficiently severe electric shock to an animal (or intruder in the case of security fences) to deter the animal from touching the fence, whilst limiting the amount of energy delivered to humans touching the fence to a level below that which will cause fibrillation effects. The relevant standard, BS EN 60335-2-76, prescribes an energy limit of 5 joules into a 500 ohm load as being safe – the 500 ohm figure being chosen to represent the typical lowest value of human body resistance at the fence operating voltages. The 5 joule figure is estimated to be below the energy needed to cause cardiac fibrillation effects in the large majority of the population.

Another example of the use of limitation of energy is found in the test probes used to carry out high voltage insulation resistance tests on cables and appliances; these probes may be energised at voltages in excess of 1 kV. The amount of current that can be delivered by the test set supplying power to the probes is commonly limited to 5 mA to ensure that anybody touching the probes will not suffer electrical injury. In situations where current, and therefore energy, limitation is used in this way, the means of limiting the current must be reliable and such that foreseeable single faults will not cause an increase in current flow.

Although not strictly associated with preventing electric shock or burn injury, a very important application of the energy limitation technique is the intrinsic safety concept used in explosion protection. The technique, designated Ex i and explained in BS EN 50020 and in Chapter 15 of this book, ensures that any electrical equipment installed inside flammable atmospheres cannot generate sufficient energy during normal operation and under fault conditions to ignite the atmosphere. This allows instrumentation to be installed in areas such as process vessels where a flammable atmosphere is likely to be present during routine operation.

Non-conducting location

Non-conducting location as a form of protection is not much used in the UK because of the practical difficulties of providing it and then ensuring its integrity is preserved. It has an application in the testing of electronic apparatus where the test area is specially designated for use by authorised personnel with the knowledge and skill for successful operation.

The essence of the concept is the prevention of direct and indirect shocks by contact between a live part or a conductive part made live by a fault and earth. To this end, insulating walls and floors are used with the minimum amount of touchable conductive and extraneous conductive parts. Any that are used and that could become live under fault conditions have to be so spaced as to prevent anyone touching two of them at the same time. The area has to be earth-free, so no protective conductors are employed and conductive and extraneous conductive parts are not earthed. Extraneous conductive parts, such as metal pipes which are in the location and outside it, need protection by insulation, barriers or placing out of reach so that in the event of a fault they cannot transmit a potential, including an earth potential, in either direction.

Earth-free local equipotential bonding

Earth-free local equipotential bonding as a form of protection is for special applications only and might be used in a test area as described in the preceding section to connect metal instrument cases and thus prevent a potential difference arising between them, which would be an indirect shock hazard.

Electrical separation

In the electrical separation technique, the supply is not earthed or otherwise referenced – such systems are commonly known as 'isolated' or 'unreferenced' supplies. The source for this type of system is usually a safety isolating transformer or its equivalent and the circuit voltage is limited to a maximum of 500 V.

The principle, as illustrated in Fig. 3.1, is that anybody simultaneously touching one pole of the unreferenced supply and earth will not experience an electric shock because there is no complete circuit back to the point of supply. However, if there were to be an insulation failure that inadvertently connected one pole to earth, the system would become an earth referenced supply and anybody touching the live pole and earth would be at risk, as illustrated in Fig. 3.1. For that reason, it is very important that unreferenced systems are routinely tested for such insulation failures. Note that Fig. 3.1

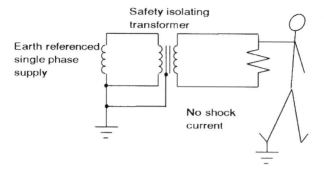

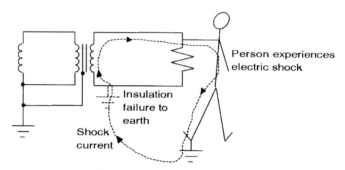

Fig. 3.1 Principle of electrical separation.

does not show the overcurrent protective devices (fuses or circuit breakers) that must be installed in both poles of such unreferenced systems.

IT supply system

The IT supply system is the same as a TT system (a TT installation is one in which the earth is derived from a local earth electrode rather than from the supplier's supply cable – see Chapter 10) but either without a source earth or with earthing through a high impedance. It is not legal for public supplies but may be used in private installations. The system is monitored to provide visual or audible warning of an earth fault. The applications include continuous processing plants and medical intensive care apparatus where the supply needs to be maintained even if an earth fault should occur. It is occasionally used to supply test facilities as it affords protection against direct shocks. As described for 'electrical separation', which is a form of IT system, a direct shock hazard arises only if there is an earth fault, and an indirect one if there are two. Earth leakage protection is provided by an RCD which interrupts the circuit on the occurrence of the second earth fault.

Earthing, equipotential bonding and automatic disconnection of the supply

Earthed equipotential bonding and automatic disconnection of supply, known as EEBADS, is the most common technique employed for protection against indirect contact electric shock. Essentially, for earth-referenced supplies, the technique requires the exposed conductive parts of Class I apparatus and equipment to be earthed by means of the protective conductor, with the protective conductor connected back to the main earthing terminal of the installation.

Extraneous conductive parts, being metalwork that does not form part of the installation but which can introduce a potential (usually earth) into the installation, must also be earthed. This is usually achieved by using bonding conductors to connect water, gas and oil pipes (and any other metallic services) to the main earthing terminal. The connection together of the exposed and extraneous conductive parts in this way creates an equipotential zone, which is normally held at earth potential. In areas of higher risk, such as bathrooms and swimming pools, supplementary bonding conductors are used to connect together all the exposed metalwork.

When an earth fault occurs, connecting a live conductor to exposed metalwork, the fault current finds its way back to the earthed point(s) of the supply system. The impedance of the fault circuit, called the earth fault loop impedance, must be low enough to ensure that the fault current is high enough to trip protective circuit breakers or blow fuses quickly enough to prevent danger. Generally speaking, socket outlet circuits and final circuits supplying handheld and portable equipment must be disconnected in 0.4 s, whereas distribution circuits and circuits supplying stationary equipment must be disconnected within 5 s. It is the amount of fault current flowing, together with the rating of the trip device, which determine the speed of disconnection. The limiting values of earth loop impedance for different types and ratings of protective devices are listed in BS 7671, which also provides the detailed requirements for EEBADS systems; see Chapter 10.

For the duration of the fault, until it is cleared by operation of the protective device, the exposed conductive parts will rise to a voltage above earth potential, determined by the magnitude of the fault current and the resistances in the fault circuit. The action of connecting the exposed and extraneous metalwork together to create the equipotential zone means that, for the duration of the fault, all the metalwork in the installation that may be at a potential will rise to the same potential, minimising the risk of electric shock injury to anybody who may be simultaneously touching different metalwork. This is the purpose of creating the equipotential zone.

In circumstances where sufficiently low values of earth loop impedance cannot be achieved, such as on TT systems, the disconnection device may need to be an RCD, as described in the next section.

The integrity of the earthing system is obviously of considerable importance in this type of system. This explains why Regulation 9 of the Electricity at Work Regulations specifically covers the integrity of the referenced conductors. Routine testing of the earthing is also important but in some circumstances, such as when there is constant flexing of the cable carrying the protective conductor, it may be necessary to take measures to increase the earthing integrity. There are techniques available for this. For example, a second protective conductor core will reduce the likelihood of a loss of earthing. Alternatively, the integrity of the earth system can be monitored by the circulating current technique.

The basic circuit of a circulating current earth monitoring system is shown in Fig. 3.2. It entails an extra core in the flexible cable. The step-down transformer's secondary winding circulates a current of a few milliamps at PELV through a contactor coil, the pilot core in the flexible cable, the protected apparatus and back to the transformer via the protective conductor. Any break in this circuit causes the contactor to open and to switch off the supply. The system is applicable to multiphase systems also. If a braided armoured flexible cable is used, the armour can be employed as the

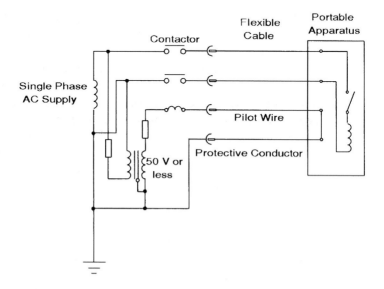

Fig. 3.2 Basic circulating current earth monitoring.

pilot conductor instead of an extra core in the cable. In this case, the armour should have an overall insulating sheath to avoid fortuitous contacts with conducting materials providing a parallel path.

This type of system is becoming less common as a result of the increasing use of reduced voltage (110 V centre-tapped-to-earth) and even battery-operated Class II hand tools in construction sites and other harsh environments. It does, however, have a role in ensuring the integrity of earthing systems where there are high levels of protective conductor current (see BS 7671, section 607).

Earth leakage protection

Circuit breakers and fuses are installed in circuits to operate in the event of excess current arising from overload conditions and faults. The most common type of fault is an earth fault, but it is frequently the case that the current flowing due to earth faults is too low to operate the overcurrent protection devices. In addition, the overcurrent protective devices will not operate in the event of somebody making direct contact with a live conductor – the current which flows through the body to earth will be too low to operate the devices but will often be high enough to cause fatal electric shocks. These two problems can be obviated by the use of earth leakage protection devices.

There are two generic types of device used for earth leakage detection: those that are voltage-operated and those that are current-operated. The voltage-operated devices are no longer used but, for completeness, they consisted of a coil connected in series in the earthing conductor or between the metalwork of the installation and an auxiliary earth electrode. The device sensed a voltage rise in the metalwork with respect to earth and, when this occurred, tripped the circuit breaker.

The current-operated devices work on a different principle, as illustrated in Fig. 3.3 for a single-phase system. When the circuits are fault-free the current flowing in the phase conductor (I_{ph}) will be the same as the current flowing in the neutral (I_n). If there is an earth fault, some current (I_{ef}) will flow back to the source via the earth path, creating an imbalance in the current flowing through the phase and neutral. It is this imbalance that is measured, usually by passing the phase and neutral conductors through a core balance transformer. Any current imbalance produces a resultant magnetic flux which is picked up by the sensing coil and which, if it reaches a predetermined level, will cause the trip coil to operate.

The current imbalance needed to operate the device varies according to the application. However, when the RCD is provided for protection against electric shock, it should have a rated residual operating current (i.e. the current imbalance that causes the device to operate) not exceeding 30 mA

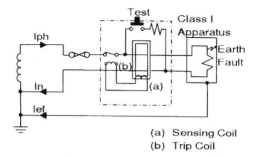

(a) Sensing Coil
(b) Trip Coil

Fig. 3.3 Single-phase RCD in simple circuit.

and an operating time not exceeding 40 ms when the residual operating current is 150 mA. Most consumer units nowadays incorporate a split in the busbars, with an integral RCD providing earth leakage protection on circuits to socket outlets.

The devices are not restricted to single-phase systems. Figure 3.4 illustrates a three-phase RCD connected into the supply from a three-phase distribution board to a motor. In this particular case, the RCD may be set to operate at a leakage current of perhaps 500 mA since it is providing protection

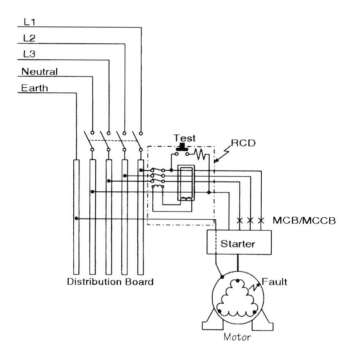

Fig. 3.4 Three-phase RCD.

against indirect contact. Note that the RCD is provided in addition to overcurrent protection devices such as miniature circuit breakers (MCB) and moulded case circuit breakers (MCCB). There are devices, known as residual current breaker with overcurrent device (RCBO) which combine the RCD and MCB functions.

Neither is the technique of earth leakage detection restricted to low voltage systems. The technique is employed on high voltage systems although the core balance method is not the only one used. For example, another way to detect earth fault current is to monitor the amount of current that flows in the earthing conductor at the point of supply, using a current transformer. If the amount of current exceeds a particular value, a circuit breaker will operate to cut off the supply.

Every residual current circuit breaker on low voltage supplies has a test button which, when pressed, creates an imbalance in the phase and neutral conductors passing through the transformer. This allows the tripping mechanism to be tested, although it does not provide a test of the magnitude of the residual operating current or the tripping time – proprietary test equipment is available for this purpose. It is very important that the test button is used periodically to confirm the RCD's serviceability because RCDs are sensitive devices and it is not uncommon for them to fail to danger; i.e. they fail in a way that means the contacts are closed but the device will not operate on demand.

This failure characteristic means that an RCD should not be relied on as the sole means of protecting against injury from direct contact. Another reason for this is that, for the RCD to operate in the event of direct contact, current of at least 30 mA must flow through the 'victim'. This amount of current is large enough to cause muscular contraction so, whereas it will almost certainly prevent electrical injury effects such as ventricular fibrillation in most cases, it may not prevent injury arising from the muscular contraction – such as falling off a ladder or being thrown against a wall. Since the Electricity at Work Regulations aim to prevent injury, and since an RCD may not prevent an injury in the event of direct contact, its use as the sole means of protection against direct contact injury would be unlikely to satisfy the law. Having said that, the device's value in providing supplementary protection against injury should not be underestimated.

There are some instances where the use of an RCD should be considered to be obligatory. These include:

■ in socket outlet circuits in TT installations;
■ in socket outlet circuits where it is foreseeable that the socket will be used to power outdoor equipment;
■ in situations where there is an increased risk due, for example, to the

presence of water; this would include the power supplies to power washers;

■ where 240 V hand tools and power tools are being used, especially in work environments such as construction sites and workshops;

■ in test areas where earth-referenced conductors may be exposed.

Many circuits and appliances generate leakage currents to earth through, for example, radio frequency filters. This means that in larger systems there can be quite a substantial amount of earth leakage current flowing through the protective conductors under normal operating conditions. In these types of installations, a 30 mA RCD installed at the origin can be subject to nuisance tripping so RCDs should be installed closer to the loads. If RCDs are installed in series, discrimination between them can be achieved by building time delays into the RCDs, with the delay highest in those RCDs closest to the point of supply. See section 607 of BS 7671 for more information on this topic.

PRECAUTIONS AGAINST BURN INJURIES

Some electrical burns occur without the victim suffering an electric shock and merit, therefore, separate preventive measures. These are described in this section.

Arc burns

Many burn injuries are caused by flame arcs emitted with explosive violence when short circuits occur in apparatus where the fault levels are high. The short circuits often take place during live working when conductive parts such as bolts, nuts and washers are accidentally dropped and bridge between a phase conductor and earthed metalwork, or between phases. Another common cause is the use of uninsulated or insufficiently insulated tools which can bridge conductors in the same way.

Whenever possible, the work should be done with the apparatus dead and so avoiding the risk, but if this is not possible then detailed preplanning of the task should be done to minimise the risk. Insulating screens between live parts of different polarity and between 'live' parts and earthed metalwork should be used and an insulating mat or stand provided for the operator. Tools should be insulated and can be magnetised to assist in the safer positioning or withdrawal of ferrous parts. Heat-resistant face shields and clothing may also be used with advantage. Insulating rubber gloves to BS 697 : 1977 are useful for some simple manipulative work but are clumsy and therefore unsuitable for more intricate tasks.

Bare instrument probes can also cause short circuits, arcing and injury to the user. Only the contact points should be bare, with a maximum length of exposed metal of 4 mm, and the rest of the probe insulated. For some applications, probes with retractable contacts can be used with advantage. Advice on the safety standards for electrical instruments for use by electricians is published in the HSE's Guidance Note GS38.

Radio frequency (RF) contact burns

A common cause of RF contact burns is from the metal electrode jigs of dielectric heating and plastics welding apparatus, operating at radio frequencies. Unless the output is pulsed there is unlikely to be a shock risk, but an accidental hand contact, for example, will draw out an arc when the hand is removed and a high frequency burn will result. The burning is a function of the arc energy, and for low power apparatus only a minor burn is likely. More powerful equipment, above about 1 kW, can inflict more serious burns.

The small welding machines usually use manual or foot power to close the jig on to the work piece and there is normally an interlocked switch to apply the RF power only when the jig is closed. As the jigs have to be changed from time to time to suit different work pieces, it is not convenient to provide guards and not really necessary as an accidental contact only causes a slight burn which operators soon learn to avoid. The larger machines have power operated jigs and the guards, necessary for the trapping risk, can be designed also to avoid the contact hazard.

Where a number of electrodes are clamped manually to the work piece, as in glue curing of joints in wood fabrications, a guard or fence can be used, interlocked with the electronic control gear to keep the operator away from the work piece when the electrodes are energised.

Radiation burns

At the frequencies allocated for industrial, scientific and medical purposes between 13.56 and 40 680 MHz, those used for heating applications at 13.56, 27.12, 896 and 2450 MHz are in general use for domestic, commercial and industrial purposes. They may employ substantial field strengths that could cause severe burn injuries, so precautions are necessary to contain the radiation and prevent access into wave guides and resonant chambers and between applicators when the apparatus is energised. This is effected by bolted-on covers which require tools for their removal, or access doors interlocked with the supply so that the supply is off when the door is open.

There is also danger from radiation leakage, so covers should be closely fitted, with metal-to-metal contact, to prevent this. Doors and other openings should be sealed with a suitable material and/or provided with wave traps. At radio frequencies, some of the conductors may act as aerials and emit radiation. If this is of sufficient power density to adversely affect nearby operators, suitable screening should be provided. This could be in the form of an interlocked guard so that access to the work piece is only possible when the RF source is off. If this is impracticable, other precautions should be taken such as the provision of remote and/or screened two-handed controls. The radiation is attenuated with distance, roughly in accordance with the inverse square law, so it is not necessary to separate the operator very far from the machine to attain safety.

High field strengths can occur in the immediate vicinity of powerful radio, radar and television aerials so barriers should be provided to exclude intruders. Operators working on or near energised apparatus should carry field strength monitors so that they can avoid undue exposure. Periodic checking of the field strength near all RF and microwave apparatus for leakage and stray fields should be carried out to ensure that safety is maintained.

Infrared radiation is immediately apparent as heat and operators instinctively move away from it. Precautions are, therefore, needed only for furnace men and the like in the form of thermally insulating or reflecting gloves, clothing and face screens. Ultraviolet radiation, however, does not cause a heat sensation and so the source should be screened or, where this is impracticable, as in the case of arc welding, the operator should cover exposed skin and view the arc through special lenses which filter out the ultraviolet and reduce the glare.

Powerful concentrated laser beams are useful for cutting and drilling and have to be guarded to prevent anyone getting into the beam path. Less powerful lasers, used for surveying for example, are not usually hazardous except perhaps to the eyes so precautions are advisable to prevent eye exposure to the beam. Lasers products are classified into four classes according to the risk of injury that they pose, and consequently the precautions that must be taken to protect people from their harmful effects; detailed information on the classes and the safety precautions is published in BS EN 60825-1:1994, IEC 60825-1:1993 Safety of laser products. Equipment classification, requirements and user's guide. The classes are:

- Class 1 Laser: devices that cannot under normal operating conditions cause a hazard.
- Class 2 Laser: low power visible lasers which, because of normal aversion responses, do not normally present a hazard, but may have some potential for injury if viewed directly for extended periods of time.
- Class 3a Laser: devices that would not normally produce a hazard if

viewed for only momentary periods with the unprotected eye. They present a hazard if viewed using collective optics.

■ Class 3b Laser: devices that can produce a hazard if viewed directly, including intrabeam and specular reflections.

■ Class 4 Laser: devices that can produce a hazard not only from direct and specular reflections, but also from a diffuse reflection. Such lasers may also produce fire hazards and skin burning hazards.

PRECAUTIONS AGAINST FIRE

An installation designed and constructed to the requirements of BS 7671, described in Chapter 10, is not a fire hazard if it is properly maintained. To minimise the danger of ignition from the several possible causes listed in the subheadings, the relevant design precautions are described.

Overloading

■ Cables should be of adequate rating for the load, having regard to environmental conditions, e.g. under or in thermal insulation or in hot environments.

■ Cables should be protected by correctly rated excess current protective devices.

■ Apparatus such as motors which may be subject to overloading should be provided with their own excess current protection and, where necessary, single phasing protection.

Earth faults

■ Provide earth leakage protection.

Short circuits

■ Ensure that circuit breakers and fuses are adequately rated for the potential fault level and can safely and rapidly interrupt the short circuit current.

Damaged wiring

The type of wiring should be suitable for the environmental conditions. Examples of ensuring this are the provision of:

- adequate protection against mechanical damage and the ingress of moisture;
- MIMS cable in very hot locations such as on furnaces;
- corrosion-resistant materials in polluted atmospheres;
- flexible conductors where there is cable movement;
- explosion-protected apparatus and wiring in flammable atmospheres.

Poor workmanship

A properly designed installation can be a fire risk if it is badly constructed. Only properly trained and competent operators should be employed and there must be adequate skilled supervision on site to ensure:

- compliance with the specification;
- a high standard of workmanship;
- that all the relevant BS 7671 tests are done and that the results are satisfactory.

Poor maintenance

Inadequate and poor maintenance enhances the fire risk so a planned fault reporting, periodic inspection, test and servicing programme is necessary to prevent the occurrence of most faults and the early discovery and rectification of those that do occur. The fault reporting system should identify faulty apparatus between inspections. The inspections should reveal such potential fire hazards as:

- loose connections;
- loose cable grips;
- dirty, misaligned or damaged contacts;
- contaminated oil in transformers, switch and control gear;
- physical damage to apparatus and wiring;
- corrosion.

The testing programme should prove the integrity of the protective and bonding conductors, that the insulation values are adequate and that the protective devices operate correctly.

Additional precautions in high fire risk locations

In timbered buildings, factories and warehouses containing flammable materials, oil depots and refineries and the like where the economic consequence of a fire merits extra precautions, the installation should be designed to minimise the fire risk consequent on the occurrence of faults. The following recommendations should be considered:

- Use air-insulated, vacuum or SF_6-insulated switchgear and control gear instead of oil-insulated equipment.
- Use air-insulated rather than compound-filled busbar chambers.
- Use metalclad switch and control gear.
- Do not use oil-insulated transformers unless they can be sited in a safe place with facilities for trapping spilt oil.
- Use a metalclad wiring system. Armoured cables should not be served and MIMS cables should not be plastics-covered.
- Wiring should be routed as far as possible clear of flammable materials and secured to non-flammable surfaces and situated where it is not vulnerable to damage.
- A fireman's switch should be provided on the outside of a building so that in the event of a fire, the installation therein can be made dead and the fire dealt with safely and expeditiously.

PRECAUTIONS AGAINST EXPLOSIONS

The precautions described in the preceding sections on burns and fires also serve to prevent the explosions consequent on short circuits in high fault level circuits. Reference should also be made to Chapter 12 for the relevant measures to avoid the explosions that can occur when an underground cable is damaged during excavation work.

Oil-immersed apparatus needs to be properly maintained if explosions are to be avoided, as referred to in the preceding section on fires, 'Poor maintenance'. Refer also to BS 6423 and BS 6626 for LV and HV switchgear and control gear maintenance and BS 5730:1979 Code of practice for maintenance of insulating oil. This code describes the deterioration that occurs in service, sampling techniques and the frequency and types of testing required. It is advisable to examine the contacts for damage and test the oil of an oil circuit breaker after it has operated to interrupt a high energy level short circuit fault.

Another precaution against the risks associated with the use of oil-filled switchgear is to replace it with switchgear that uses an alternative insulating medium. Options include sulphur hexafluoride (SF_6) gas-filled enclosures, air insulation, and vacuum circuit breakers. The SF_6 variety is beginning to lose favour because of the environmental and toxic hazards that the gas poses, so vacuum and air-insulated gear is now preferred by most high voltage users and distributors. Another advantage that modern high voltage switchgear has over its predecessor designs is that it is mostly of the fixed pattern without having withdrawable components. This means that features such as busbar and cable spout shutters are not needed, with a consequential increase in safety arising from the fact that engineers operating the switchgear do not need to work close to uninsulated conductors.

Electrical equipment installed in flammable or dust-laden explosible atmospheres needs to be selected so that it cannot act as an ignition source. This requires hazardous areas to be classified according to the likelihood of a flammable or explosible atmosphere being present, and then suitable equipment being selected, installed and maintained by competent persons. Detailed information on the requirements is contained in Chapter 15.

SAFE SYSTEMS OF WORK

The safety precautions described so far have concentrated on hardware solutions. However, it should be recognised that a large proportion of accidents result from people adopting unsafe systems of work on electrical systems. It is therefore clear that the adoption of safe systems of work should have clear benefits in accident reduction. Safe systems of work are frequently referred to as 'software' measures, in comparison with the hardware measures already covered.

The best way that employers can put in place suitable safety management procedures covering safe systems of work on electrical systems is to produce a set of electrical safety rules. These rules, which should be drawn up by somebody familiar with the electrical systems at the premises, the nature of the work being done on them, and the principles of electrical safety, should explain the employers' policies for achieving safety from the electrical system. In reality, there are two main areas to be considered: precautions on equipment that has been made dead, and precautions during live working. These procedures, which are explained below, should be based on the outcome of a risk assessment carried out before activities are undertaken. The legal duties for risk assessments are explained in the chapters dealing with legal requirements but, in essence, a risk assessment is a qualitative judgement on the level of risk associated with a particular activity, and a determination of the measures which need to be taken to control the risks.

A risk assessment should identify the hazards arising from the activity, their severity (usually expressed as high, medium, or low), the likelihood of the hazard occurring, and the likelihood of being able to avoid the hazard. These factors then need to be combined to derive an overall assessment of the risk. Having done that, those activities that are judged to have unacceptably high levels of risk must have action taken to bring the risk down to a tolerable level, and this will usually result in some form of written method statement or procedure which must be adopted by the personnel involved. Risk assessments are usually undertaken by teams of people, often under the control of a health and safety adviser; it is important that the people who will actually be undertaking the work should participate in the risk assessment so that they have a sense of ownership of the control measures and are therefore more likely to implement them.

Safe isolation procedures

It should be the norm that work is carried out on systems that have been made dead and on which precautions have been taken to prevent them being reenergised while work is going on. The generic safe isolation procedure is as follows:

(1) Identify the circuit or apparatus on which work is to be done.
(2) Identify all possible points of supply.
(3) Disconnect the supply by, for example, switching off an isolator, withdrawing a plug, or tripping to 'off' a circuit breaker that has a contact gap that can positively be seen to be open and which has a gap large enough for isolation purposes.
(4) Secure the point of isolation by applying a padlock with a unique key, or by locking an enclosure door, or by any other equally effective manner. As a minimum if it is not possible to lock off the means of isolation, apply warning tape to the device used to achieve the disconnection, whilst recognising that this may not be secure enough if it is at all foreseeable that the tape may be removed or ignored.
(5) Post caution notices at the point(s) of isolation to warn that work is being done on the disconnected circuit(s).
(6) At the point of work, prove the conductors are dead using a suitable voltage indicator such as test lamps. The voltage indicator must be proven to be serviceable immediately before and after the conductors have been tested for voltage. A proving unit should be used for this, although another live circuit may be used.
(7) In the cases of high voltage equipment, and some higher risk low voltage systems, applying earths to the conductors on which work is to be done enhances safety. This is to ensure that all dangerous electrical energy is dissipated and to maintain the conductors at earth potential.
(8) Post danger notices on adjacent live equipment and circuits.
(9) Ensure that the persons carrying out the work on the isolated systems are aware of the scope and limitations of the work to be done.

Safe isolation procedures such as these should only be carried out by people, such as electricians and technicians, who have been trained to implement them and who are familiar with the equipment involved. See Chapter 14 for more information on competence and authorisation procedures.

In the cases of high voltage systems and high fault level low voltage systems, it is accepted practice that the safe isolation procedures should be formalised through 'permits to work'. Essentially, an electrical permit to work is a document, normally over two sides of A4 paper, which describes the work to be done and identifies the equipment to be worked on. It lists the points of isolation and identifies where any earths and caution/danger notices have been applied. It is good practice to attach to the permit a switching schedule that identifies the sequence of actions needed to make the system safe. The permit is signed and issued by an authorised person who has actually carried out or supervised the isolation procedures. It is signed and received by the person who is going to be working on the equipment – these may be one and the same person. The permit is cancelled once the work has been completed and the system restored.

Note that electrical permits to work should only be issued for work on systems that have been made safe. They should not be used to authorise live work since, by definition, live working is carried out on systems that have not been made safe.

Model permits to work are published in the HSE's Guidance Note HS(G)85 Electricity at work – Safe working practices, and in BS 6626.

Live working practices

Live working is frequently carried out during fault finding and testing activity, particularly on low voltage systems. In many cases it is unnecessary because the work could be done in other ways with the equipment dead but, nonetheless, there is no doubt that there are many instances where live work is justified. Some specialised activity, such as live cable jointing and phasing out high voltage conductors, is carried out by specialists who have received in-depth training, usually in the electricity supply industry. The main hazard considered here is direct contact with live conductors and the aim of the precautions to be taken is to prevent injury from such direct contact.

The most important precaution is to ensure that the people carrying out live work are competent for the task or are being closely supervised by somebody who is competent. This means that they must have been trained in the task and be assessed as being competent, must understand the system on which they are working, and must be provided with the appropriate tools, test equipment and personal protective equipment. Live work should only be undertaken after a risk assessment has been conducted, the risks identified and the appropriate safety precautions have been determined. The types of precautions that can be taken are as follows:

(1) Ensure that the area of the work activity is barriered off or otherwise delineated and protected to prevent the workers being distracted or disturbed. This will also ensure that non-competent or unauthorised personnel are kept away from the live conductors. Danger notices can be used to highlight the danger from the exposed conductors. In permanent test areas it is common practice to have flashing warning lights to highlight when testing work is being done.

(2) Use tools and test equipment suitable for the job. Suitably rated insulated tools such as screwdrivers and spanners should be used. Test equipment should conform to the guidance in the HSE's Guidance Note GS38 on electrical test equipment. This means, for example, that probes should be fused, should have finger guards to stop the fingers slipping down the probe on to the live conductors, and there should only be 2–4 mm of metal exposed at the tips of the probes. Care should be taken to ensure that the test equipment itself, such as Class I oscilloscopes and signal generators, do not introduce additional hazards.

(3) Where necessary, shroud off metalwork in the vicinity of the work that may be at other potentials, including earth. For example, when live work is being carried out inside an equipment enclosure or panel, the metal structure of the panel will be at earth potential – it should be shrouded to reduce the possibility of a live-to-earth shock being experienced. This can be done using flexible insulating sheeting made from materials such as neoprene and polythene. If the floor is conducting, use insulating rubber mats to remove the shock path to earth.

(4) Where it will contribute to risk reduction, arrange for the live worker to be accompanied. The benefits are two-fold; firstly, the second person may be able to offer a second opinion or to detect unsafe practices before they lead to injury; and, secondly, he will be able to take emergency action and render first aid if things do go wrong. Accompaniment is not always essential but it should be considered.

(5) Use appropriate personal protective equipment. Antiflash clothing and eye protection will be appropriate where there is a risk of flashover or arcing. Insulating rubber gloves will be appropriate when live conductors are being handled. The PPE must be maintained in good condition.

Chapter 4

The Legislative Framework

INTRODUCTION

Criminal law addressing electrical safety is spread across a number of Acts and Regulations put before Parliament by different Government departments, or Secretaries of State, with different 'audiences' in mind. The target audiences are, generally speaking, those who supply electrical equipment for work and/or domestic use; those who use electrical equipment at work; and those who generate, transmit and distribute electrical energy. The relevant Regulations are explained in the following chapters.

Some of the legislation, such as the Electricity at Work Regulations 1989, is made under the Health and Safety at Work etc. Act 1974 and is aimed at ensuring the safety of employees, the self-employed and those who may be affected by work activity. Other legislation, such as the Electrical Equipment (Safety) Regulations 1994, is derived from European Directives (Article 95 Directives) whose main aim is to ensure the free movement of goods throughout the European Union, but which have a subsidiary but very important safety content; the Department of Trade and Industry usually takes the lead on the development of this type of legislation. Other legislation, again derived from European Directives (Article 137 Directives), is targeted at worker safety, an example being the Provision and Use of Work Equipment Regulations 1998; the Health and Safety Commission usually takes the lead on the development of this type of legislation. The Electricity Supply Regulations are the responsibility of the DTI and are aimed at ensuring that suppliers of electricity meet certain performance standards, including those covering electrical safety.

GOAL-SETTING AND PRESCRIPTION

A feature of modern UK-originated legislation in the field of health and safety is that it tends not to be prescriptive in describing what duty holders need to do to comply with the law. It is more goal-setting, describing in broad terms what needs to be achieved, but not how to achieve it. Law derived from European Directives tends to be slightly more prescriptive, as exemplified by the detailed Essential Health and Safety Requirements that form part of the Machinery Directive and its enactment in the UK, the Supply of Machinery (Safety) Regulations.

The lack of prescription in the legislation is meant to be offset by the provision of comprehensive guidance and the setting of benchmark standards by bodies such as the Health and Safety Commission/Health and Safety Executive (HSC/HSE) and the British Standards Institution (BSI). Trade associations and other similar organisations have an important role in this respect, and guidance material published by bodies such as Electrical Equipment Manufacturers and Users Association (EEMUA), the Institution of Electrical Engineers (IEE), the Electrical Contractors' Association (ECA) and Electrical Contractors' Association of Scotland (SELECT) is extremely useful.

One means at the disposal of the HSC to be more prescriptive is the development of an Approved Code of Practice (ACOP) to support a set of regulations. There is, for example, an ACOP supporting the Provision and Use of Work Equipment Regulations 1998 and one was published in support of the Electricity at Work Regulations to cover the use of electricity in quarries. An ACOP is something of a peculiar artefact because it has the status of neither a regulation nor guidance. The Health and Safety at Work Act, section 17, stipulates that failure to follow an ACOP provision is not an offence in itself. However, where criminal proceedings allege the breach of a particular regulation for which there is a relevant provision in an ACOP, and it can be proved that the ACOP provision was not followed, then this can be taken as proof of the contravention of the regulation. The defendant may escape a guilty determination, however, if he can prove to the satisfaction of the court that he took alternative steps that had the same level of safety as the ACOP's provisions. Note that provisions set out in guidance material or British Standards do not enjoy this level of legal status.

It has to be said that the goal-setting nature of the legislation causes problems for hard-pressed managers in small and medium size enterprises who do not have ready access to health and safety professionals – many of them would much rather be told precisely what they need to do to comply with the law than have to spend time and energy searching for and interpreting guidance and standards. There is an overriding need for simple, straightforward, and unambiguous guidance for these organisations, and the HSC/HSE in particular is working hard to meet the need by publishing its series of 'Essentials' guidance material.

REASONABLE PRACTICABILITY

A fundamentally important concept embedded in much of the health and safety legislation is that of reasonable practicability. Many of the regulations require action to be taken in the context of 'so far as is reasonably practicable', often spoken or written about in terms of reducing risks to a level that is as low as reasonably practicable (the so-called ALARP principle). This means that there should be a balance between, on the one hand, the level of risk reduction being sought, against on the other hand, the cost in terms of money, time or trouble needed to reduce the risk. So, for example, if there is a disproportionate cost needed to minimise still further an insignificant risk, the further risk reduction measures could be judged not to be reasonably practicable.

In general terms, any company or other entity defending a criminal health and safety charge needs to be able to prove that it was not reasonably practicable to do more than was in fact done or that there was no better practicable means than was in fact used. Having said that, the defendant only needs to be able to prove the point on the balance of probabilities, rather than using the more onerous 'beyond reasonable doubt' test that applies to the prosecution.

The use of the term 'so far as is reasonably practicable' implies that an assessment of the risk must be carried out before a judgement can be formed on whether or not a measure is reasonably practicable. Whereas the need for a risk assessment is not explicitly laid out in electrical safety legislation, it is explicitly a requirement of the Management of Health and Safety at Work Regulations. So risk assessment practices and procedures are a cornerstone of electrical safety management practice.

Some regulations do not have the 'so far as is reasonably practicable' caveat. Many of these regulations have an absolute duty for compliance; for example, in the Electricity at Work Regulations, Regulation 10 requires that every joint and connection in a system shall be mechanically and electrically suitable for use. In cases such as this, a person charged with an offence under the Regulation has the defence of due diligence available.

ENFORCEMENT

The different items of legislation are enforced by a variety of agencies, including HSE inspectors of health and safety; DTI engineering inspectors; and local authority environmental health officers (EHOs) and trading standards officers. Which agency enforces which legislation depends on the particular legislation and the location or type of premise in which the electrical system is installed or is operating. As examples:

- The DTI's engineering inspectors enforce the Electricity Supply Regulations.
- Health and safety legislation is enforced in factories, hospitals, construction sites, nuclear and offshore sites, railways, local authority premises and so on by HSE inspectors.
- Health and safety legislation is enforced in shops and offices and similar locations by EHOs.
- Trading standards officers enforce the Electrical Equipment (Safety) Regulations for electrical equipment used in the domestic market, whereas the Regulations are enforced by HSE inspectors for equipment used in premises for which HSE has enforcement responsibilities.

Enforcement action may mean different things for the different agencies. For example, HSE inspectors and EHOs exercising their powers under the Health and Safety at Work Act 1974 have considerable discretionary powers. For example, they are empowered to:

- Enter work premises at any reasonable time.
- Make examinations and investigations, including taking measurements and photographs, as necessary in the exercise of their powers.
- Serve improvement notices requiring action to be taken within a prescribed timescale. This type of notice can be appealed at an Industrial Tribunal.
- Serve prohibition notices requiring immediate action to obviate an imminent risk. This type of notice can be appealed at an Industrial Tribunal.
- Take dangerous articles or substances into possession and have them examined and tested.
- Arrange for samples to be taken and tested.
- Take statements from witnesses. Anyone suspected of breaching a legal requirement must be cautioned before being interviewed.
- Take prosecutions under the relevant legislation. In Scotland, the decision on whether or not to prosecute is taken by a Procurator Fiscal, not by the HSE inspector or EHO.

Most other enforcement agencies do not have such wide ranging and extensive enforcement powers.

PENALTIES

Individuals or companies found guilty of criminal charges under the legislation are liable to various penalties. If the case is taken under a section of the Health and Safety at Work Act, for example, and if it is heard using summary proceedings in a magistrate's or sheriff's court, the maximum penalty is presently set at £20,000. If the case is heard on indictment, in front of a jury, the maximum penalty is imprisonment for a term not exceeding two years, or a fine, or both. If the case is taken under one of the Regulations and is heard using summary proceedings in a magistrate's or sheriff's court, the maximum penalty is presently set at £5000. If the case were heard on indictment, in front of a jury, the maximum penalty is imprisonment for a term not exceeding two years, or a fine, or both.

Chapter 5
The Health and Safety at Work etc. Act 1974

INTRODUCTION

The Health and Safety at Work etc. Act 1974 (HSW Act) does not contain any provisions relating specifically to electrical safety. However, it is an Act that has had, and continues to have, an immense impact on the UK's health and safety system, effectively acting as its backbone for the past 27 years. It therefore deserves a chapter in this book to outline the main duties imposed on those at work. It is not my intention, however, to cover the many other provisions of the Act. The most important features of the Act are:

■ It is an enabling Act, allowing the promulgation of subsidiary regulations.
■ It sets goal-setting, non-prescriptive duties on employers, the self-employed and employees to secure the health, safety and welfare of those at work and the health and safety of people, such as members of the public, who may be affected by work activity.
■ It covers all work activities with the sole exception of those employed as domestic servants in private households.
■ It established, and defined the general functions of, the Health and Safety Commission as a corporate body comprising representatives of employers, trades unions, consumer bodies and so on, appointed by the Secretary of State. It also established and defined the general functions of the Health and Safety Executive.
■ It set out the enforcement powers of health and safety inspectors, as discussed in Chapter 4.

MAIN DUTIES

Section 2

Section 2 of the Act sets out the general duties that employers have to their employees for their health, safety and welfare. The flavour can, perhaps, best be understood by considering the duty in section 2(1) on every employer to ensure, so far as is reasonably practicable, the health, safety and welfare at work of all his employees. The very general nature of this duty reinforces the broad scope of the Act and its non-prescriptive character. It also points to the need for considerable discretion by the enforcing authorities when interpreting such a generic duty.

The rest of section 2 puts a bit more flesh on the general duty expressed in section 2(1). Important requirements in the context of electrical safety include the provision of plant and systems of work that are safe; the provision of information, instruction, training, and supervision as necessary; the maintenance of 'places of work' in a safe condition; and the provision and maintenance of a safe and healthy working environment.

Section 3

Section 3 deals with the employer's and the self-employed's duties to persons other than their employees. Essentially, their work activities should not expose persons not in their employment, and others, to risks to their health and safety. They also have a duty to provide others with information about work activity which might affect their health and safety. An example would be a live cable jointing team working in an excavation in a public road – the team would need to ensure that passers-by could not fall into the excavation or otherwise come into contact with the exposed live conductors, perhaps by erecting barriers with suitable warning notices affixed. Alternatively, an electrical contractor who installs a new circuit in a domestic property, but leaves it in a dangerous condition to the risk of the householder, would be in breach of section 3.

Section 4

Section 4 applies to persons in charge of premises where people work and who are not their employees. They must ensure that the premises, any plant (including the fixed electrical installation) and the means of ingress and egress are safely maintained. An obvious example of the application of this duty is a construction site where the main contractor is the person in charge and in control of the site and who is responsible, therefore, for ensuring that the site is safe for the subcontractors' labour forces.

Section 6

Section 6 applies to manufacturers, designers, importers and suppliers of articles and substances for use at work. In the context of articles, which would include electrical equipment and apparatus, the principle duties are that they must ensure that the articles are safe when being set, used, cleaned or maintained by a person at work. They must also ensure that operating and maintenance instructions are available, and must provide revisions of information when it becomes known that something gives rise to a serious risk to health and safety.

The latter requirement is commonly used to ensure that users of equipment are informed when, after an accident investigation has been carried out, it becomes apparent that the equipment needs to be modified in some way to prevent a recurrence of the incident.

The section entitles a supplier to rely on the safety-related research and testing already done by others. He does not have to repeat it. For example, an electrical contractor who installs electrical equipment such as a circuit breaker can relay on the assurances of the device's manufacturer that it has been made and tested to an appropriate British Standard. The contractor, however, retains responsibility for selecting the correct type of circuit breaker for the application ensuring, for example, that it has adequate breaking capacity for the prospective fault current. However, importers of foreign apparatus must satisfy themselves that the designer's or maker's assurances are valid.

In the event of an incident, it is a defence to prove that the occurrence could not reasonably be foreseen. The argument would be that the scientific and technical knowledge available at the time that the product was marketed was not such as might be expected to reveal the deficiency.

Section 7

Section 7 imposes duties of care on employees for their own and other's safety. They have to cooperate with their employer and anyone else to fulfil their obligations. An example would be when an electrician ignores the instructions and procedures put in place by his employer to work on an electrical installation once it has been isolated and made safe and, instead, carries out the work live. Such behaviour would probably be a contravention of section 7 and it is not unusual for cases to be brought before the courts when employees behave in this way and where somebody is harmed as a result.

Section 8

Section 8 prohibits intentional or reckless interference or misuse of anything provided in the interests of health, safety or welfare.

Chapter 6

The Electricity at Work Regulations 1989

INTRODUCTION

The Electricity at Work Regulations 1989 are made under the Health and Safety at Work Act. They were promulgated on 1 April 1989 and replaced the long-standing previous regulations on electrical safety, the Electricity (Factories Act) Special Regulations 1908 and 1944. These latter Regulations had been drafted by J. Scott Ram, who had been appointed in 1902 as the first HM Electrical Inspector of Factories and who, with remarkable pre-science, laid down the basic electrical safety principles that persist to this day.

Unlike the older 1908 and 1944 Regulations, which applied only to factory premises covered by the Factories Act, the Electricity at Work Regulations 1989 apply to all workplaces covered by the HSW Act. Their publication in 1989 was therefore the first time that detailed electrical safety statutory requirements had been applied to non-factory premises.

The Regulations are enforced by the HSE's inspectors of health and safety and by local authority environmental health officers. The HSE's general regulatory inspectors are supported by field-based inspectors who are professionally qualified and experienced electrical engineers. These electrical inspectors work in specialist groups that are part of HSE's Field Operations Directorate; there are seven such groups distributed across the UK. The electrical inspectors cover all aspects of electrical engineering, including safety-related control systems, and frequently appear in court and coroners' courts to provide expert evidence. There are other electrical inspectors based in HSE's Technology Division, who are responsible for developing technical policy.

The Regulations are supported by an HSE Memorandum of Guidance, referenced HS(R)25, and there are Approved Codes of Practice covering electrical safety in mines and in quarries.

THE REGULATIONS

The Regulations are promulgated in four Parts, with Parts I, II and IV having general application and Part III applying only to mines. In coming to terms with gaining an understanding of the Regulations, it is helpful to consider them as having two main blocks: those that deal with electrical hardware (Regulations 4(1), 4(4), 5, 6, 7, 8, 9, 10, 11, 12 and 15) and those that deal with so-called 'software', or systems of work, requirements (Regulations 4(2), 4(3), 13, 14 and 16). This distinction is useful because it recognises the fact that high standards of electrical safety will only be achieved if electrical installations and equipment meet appropriate build standards, and if persons at work on electrical systems adopt safe systems of work.

The following paragraphs give a brief explanation of the main issues covered by each of the Regulations, with the text of the main block of Regulations in Parts II and IV being spelled out.

Part I: Introduction

Regulation 2 – Interpretation

Regulation 2(1) provides interpretations of words used in the rest of the Regulations. 'Danger' is defined as the 'risk of injury', where the specific injuries covered by the Regulations are listed as death or personal injury from electric shock, electric burn, electrical explosion or arcing, or from fire or explosion initiated by electrical energy. Note that other injuries, such as crushing or shearing injuries resulting from an electrical fault in a machine's control system, are not included. The risk of injury arising from control system faults on machinery is more in the purview of the Provision and Use of Work Equipment Regulations than the Electricity at Work Regulations.

'Electrical equipment' includes equipment used to generate, provide, transmit, transform, rectify, convert, conduct, distribute, control, store, measure or use electrical energy. The reach of the Regulations is therefore very wide and all-embracing.

Regulation 3 – Persons on whom duties are imposed by these Regulations

This Regulation places the duty to comply with the Regulations on employers, the self-employed, and mine managers and operators of quarries, for matters which are within their control. The Regulation was updated in 2000 to take account of the provisions of the Quarries Regulations 1999, which introduced quarry 'operators' as legal entities.

The Regulation refers to matters which are within the control of people. This would mean, for example, that if a company were to employ an electrical contracting company to carry out work on their electrical installation, and if they had taken all reasonable steps to ensure the competence of the contractors, but an electrician employed by the contractors were to suffer an electrical injury, the company would have the defence that they were not in control of the electrical work.

Regulation 3(2) mirrors the duty that the HSW Act, section 7, places on employees by requiring them to cooperate with their employers and to comply with the Regulations.

Part II: General

Regulation 4 – Systems, work activities and protective equipment

'(1)   All systems shall at all times be of such construction as to prevent, so far as is reasonably practicable, danger.

(2)   As may be necessary to prevent danger, all systems shall be maintained so as to prevent, so far as is reasonably practicable, such danger.

(3)   Every work activity, including operation, use and maintenance of a system and work near a system, shall be carried out in such a manner as not to give rise, so far as is reasonably practicable, to danger.

(4)   Any equipment provided under these Regulations for the purpose of protecting persons at work on or near electrical equipment shall be suitable for the use for which it is provided, be maintained in a condition suitable for that use, and be properly maintained.'

Regulation 4 is an 'omnibus' Regulation which is fleshed out by subsequent Regulations. The four subsidiary Regulations are dealt with separately in the following text.

Regulation 4(1)

Regulation 4(1) simply says that all systems must be safe, so far as is reasonably practicable. The main means of complying with the requirement is to ensure that electrical equipment, apparatus and installations are designed by a competent person to a recognised standard. As far as low voltage distribution systems are concerned, the benchmark standard is BS 7671 : 2001 Requirements for electrical installations – the IEE Wiring Regulations 16th Edition. Indeed, the Memorandum of Guidance to the Regulations specifically comments that systems built and installed in compliance with that standard are likely to achieve compliance with the Regulations. Of course, it is axiomatic that the system would need to be installed by a competent person and tested to prove its safety before being commissioned and put into use.

There are many other standards that offer guidance on the safe design of electrical systems and equipment. It is impracticable to list them all here but two are worthy of mention. BS EN 60204 Safety of machinery – electrical parts of machinery, is a CEN/CENELEC Type B standard (see Chapter 13) with comprehensive guidance on electrical safety principles relating to machinery. The BS EN 60335 series of standards provides equally comprehensive guidance relating to electrical safety in commercial and domestic equipment such as food mixers, meat slicers, white goods, fly 'electrocutors' and so on.

Regulation 4(2)

Regulation 4(2) requires that systems are maintained to prevent danger, but only where maintenance is necessary to prevent such danger; it has the 'so far as is reasonably practicable' caveat. Note that there is no prescription here on the form that the maintenance should take, and that there is no duty for records of the maintenance to be kept. However, it is generally recommended that records are kept as a means of demonstrating compliance to enforcement authorities and as a means of detecting deterioration over time.

It should also be recognised that the Regulation makes no mention of portable appliances, despite the widespread belief that it is only portable appliances that need to be inspected and tested. This belief is reflected in the fact that many organisations have comprehensive and expensive inspection and test programmes for portable appliances (kettles, computers, radios and the like) but ignore the need to maintain their fixed distribution systems and equipment. Indeed, largely at the behest of companies that offer portable appliance testing services, it has to be said that many organisations over-maintain their portable appliances by carrying out testing too frequently. The uncertainties associated with the maintenance of electrical equipment led the HSE to issue a clarifying guidance note on the topic, HS(G)107 Maintaining portable and transportable electrical equipment.

Maintenance of electrical systems generally consists of a mixture of inspections and tests. Low voltage installations should be maintained following the advice contained in BS 7671 : 2001, which covers the appropriate visual inspections and periodic tests (see Chapter 17 for a description of the inspections and tests). The IEE's guidance note on inspection and testing is also a valuable source of advice, including recommendations on the periods between inspections and tests.

Advice on the maintenance of high voltage switchgear is published in BS 6626.

As far as other equipment is concerned, it is often best to follow the advice of manufacturers and suppliers on the form of preventive maintenance needed. In the absence of such advice, the IEE's excellent guidance note, Code of practice for the in-service inspection and testing of electrical equipment, provides comprehensive guidance. This document correctly emphasises the importance of routine visual examinations of equipment for the very good reason that most faults, such as damaged insulation on cables, which lead to danger can be detected by a simple visual inspection. This includes a before-use inspection by the operator of the equipment as well as periodic inspections by a competent person, who does not necessarily need to be an electrically-qualified person such as an electrician.

Regulation 4(3)

Regulation 4(3) requires that any work activity on or near a system is carried out safely, again with the 'so far as is reasonably practicable' caveat. In this context, the caveat strongly implies that a risk assessment should be done before the work commences. Whereas this is not explicitly spelled out, it is made a specific requirement by the Management of Health and Safety at Work Regulations which embrace all work activity, including electrical work.

The work activity includes operation, use and maintenance. Those responsible for the work, and those engaged in it, need to be competent to appreciate the electrical hazards and the control measures needed to mini-mise the risks to an acceptable level; these issues are addressed in Regulations 14 and 16. To this end, the person responsible for managing the work should consider the work activities, the risks that arise, the measures that should be taken to control them, and who should be allowed to perform them. If, for example, work has to be done on a low voltage system which necessitates it being made 'dead', the supply to it needs to be isolated, the isolator locked off and a test made to prove that the isolated part is indeed 'dead' and therefore safe to work on. Anyone authorised to carry out these safety precautions has to be familiar with the system, knowing which isolator or isolators to open and lock off and how to apply the test. The responsible person should also ensure that those who do the work have appropriate technical knowledge to do it properly and, on completion, to test it to prove its safety.

Where work has to be done near a system where there may be danger, the responsible person has to provide for the safety of the workers. For example, painters, who may not be electrically knowledgeable, may have to decorate an overhead crane track where there are bare live trolley wires. The painters must be warned about the hazard, and measures must be taken to avoid the danger. This might involve the provision of track stops to limit crane movement, and the painters working in a segregated area where the live conductors are screened to prevent direct contact. Where systems are concealed, however, the danger may be less obvious, but measures still need to be taken. Buried cables are a potential hazard during excavation work so their location must be ascertained before the work starts and the workers instructed on the precautions to be taken to avoid the cables being damaged in a way that may result in injury.

Where live work is to be carried out, such as live jointing work or fault-finding on live equipment, the risk of electrical injury tends to be high, so careful planning is needed. Such work must only be done by competent people who have the appropriate skills and knowledge. The types of precautions that can be taken were explained in Chapter 3 but, in brief, are:

- Ensure that the area of the work activity is barriered off or otherwise delineated and protected.
- Use tools and test equipment suitable for the job.
- Where necessary, shroud off metalwork in the vicinity of the work that may be at other potentials, including earth potential.
- Where it will contribute to risk reduction, arrange for the live worker to be accompanied.
- Use appropriate personal protective equipment.

An uninstructed person can safely carry out many operation and use activities such as operating a light switch. However, dangerous use often stems from abuse and/or ignorance on the part of the user. The careless operator of a portable angle grinder, for example, might abrade the flexible cord and expose a live conductor, resulting in a potential shock hazard to anyone who might touch it. Again, if the user of a machine exceeds its duty cycle by continuous use on overload, it may overheat, the insulation may fail and the metal carcass become live, endangering the user. So it is important that those who are engaged in operation and use activities are trained to avoid danger.

Regulation 4(4)

Regulation 4(4) deals with protective equipment such as permanent and portable insulating stands or screens, insulating boots and gloves used to prevent direct contact injuries, insulated tools, and antiflash clothing used to protect against flashover burn injuries. In fact, the reach of the Regulation is wide and would include, for example, the fencing of an overhead power line crossing on a construction site to prevent plant or scaffold poles making contact with the line. It would also include door interlocks on control panels used to prevent access to live apparatus, insulated tools and potential indicators.

Many employers require their electrical staff to provide their own tools for use at work, including insulated tools for use during live work. Whereas this may be standard practice, and the tools may be the personal property of the employees, it does not absolve the employer from ensuring that the tools are suitable for the work being carried out by the electricians and technicians. Employers are well-advised to carry out periodic inspections of their employees' tools to ensure that they are suitable and in a safe condition, and are maintained as such.

Note that this Regulation does not attract the 'so far as is reasonably practicable' caveat – it is an absolute duty that equipment used for protective purposes has to be suitable, and properly maintained and used.

Regulation 5 – Strength and capability of electrical equipment

> 'No electrical equipment shall be put into use where its strength and capability may be exceeded in such a way as may give rise to danger.'

Regulation 5 refers to the ability of equipment to withstand the effects of the currents and voltages to which it is likely to be subjected. These include the stresses that will occur when operating at its normal rating, on overload and under fault conditions and when subjected to mains-borne transients. The key to compliance is the selection of the right equipment by a competent person. For example, in choosing a direct-to-line motor starter, one which complies with BS EN 60947 is preferable because its performance details are ascertainable and because it is tested to the standard. Obviously, it should be rated to match the full load, starting and stalled currents of the induction motor it controls and, also, the duty cycle. Inching requirements and the number of starts per hour need to be known, and also any adverse environmental conditions that may exist.

The next consideration would be fault protection. If a short circuit fault occurred on the cable between the starter and the motor, is the starter capable of safely interrupting the fault current? A loop impedance test may be needed to ascertain the fault level. If the chosen starter is incapable of safely interrupting the fault current, suitable additional protection would be needed on the supply side of the starter. This would ensure that the operation of the combined excess current protective devices would either avoid damage to the starter or so limit it that anyone nearby is not endangered.

Another example is the selection and use of suitably insulated cables. Their voltage and current ratings should not be exceeded where this might cause an insulation failure and the consequential exposure of a live conductor that would produce the risk of electric shock and burn injuries to anyone who may touch it. Yet another example is high voltage switchgear, such as circuit breakers. These devices may be required to clear short circuit faults on

circuits where the fault level is extremely high, normally up to 250 MVA on
11 kV distribution circuits, for example. If the circuit breaker's rated fault
breaking capacity were to be less than the fault level on the system, short
circuit faults would have the potential to cause catastrophic failure of the
switchgear, resulting in an explosion or fire.

It is not unknown for circuit breakers which are rated at, say, 50 MVA,
and which were installed on a network in the 1940s and 1950s when they were
adequately rated, to become underrated because system reinforcements have
resulted in increases in the fault level above 50 MVA. Another cause of
devices becoming dangerously underrated stems from increases in the
quantity and power rating of rotating plant in the load centres fed from a
circuit breaker – the contribution these devices make to fault level calculations is frequently overlooked.

Regulation 6 – Adverse or hazardous environments

'Electrical equipment which may reasonably foreseeably be exposed to –
(a)   mechanical damage;
(b)   the effects of the weather, natural hazards, temperature or pressure;
(c)   the effects of wet, dirty, dusty or corrosive conditions; or
(d)   any flammable or explosive substance, including dusts, vapours or gases,
      shall be of such construction or as necessary protected as to prevent, so far
      as is reasonably practicable, danger arising from such exposure.'

Regulation 6 has the important stipulation that measures must be taken to
prevent dangerous deterioration of equipment that may be exposed to conditions that may adversely affect its performance. The specific influences
considered are mechanical damage; the effects of the weather, natural
hazards, temperature or pressure; the effects of wet, dirty, dusty or corrosive
conditions; and any flammable or explosive substances, including dusts,
vapours or gases.

Compliance requires the selection of appropriate equipment for the pre-
vailing conditions, and those conditions that can be foreseen for the location
where the equipment is to be used. Alternatively, it is sometimes the case that
the equipment can be otherwise protected to achieve compliance, such as by
placing it out of harm's way in the case of potential mechanical damage to
cable runs.

The most common protection against mechanical damage is making the
construction of the equipment sufficiently robust for its likely applications.
For example, portable equipment used on construction sites will need to be
able to withstand the very rough treatment commonly meted out by
personnel who may not be very aware of the electrical risks. In contrast,
electrical equipment designed for domestic use will usually not need to be

quite so well protected against mechanical damage. BS EN 50102:1995 defines an IK code to deal with the degrees of protection provided by enclosures of electrical equipment against external mechanical impacts. The code ranges from IK00 to IK10, with increasing levels of impact energy.

The IP code associated with protecting enclosures against the ingress of liquids and dusts has been described in Chapter 3. This has important application in securing compliance with Regulation 6.

Lightning is a weather hazard that is often overlooked. The 30% of electricity supply failures on overhead line transmission and distribution systems that are due to lightning strikes demonstrate the extent of the risk and the consequential need for the owners to provide protection, so far as is reasonably practicable, to avoid danger from the strikes. Section 3 of the HSW Act also requires employers to conduct their undertakings in such a way as to avoid risks to persons not in their employment. On public supply systems, some attempt is made to protect the high voltage (HV) system but nothing is usually done to protect the low voltage (LV) overhead distribution system. The consequence is that powerful, high voltage transients can damage the service line within the consumer's premises and, sometimes, his installation. This can result in fire and may result in persons in the premises being injured from electric shock and/or burn injuries from electrical equipment damaged by the lightning impulse.

A range of British Standards extensively covers the prevention of explosions caused by electrical apparatus acting as ignition sources in flammable and dust-laden atmospheres. These give the best route to securing compliance with the requirements of Regulation 6 covering flammable atmospheres. For example, BS EN 60079-10 is a harmonised European Code of Practice published to replace BS 5345, which had been the standard covering UK electrical installations in potentially flammable gas atmospheres since 1976. The standard provides guidance on zone classification of hazardous areas. BS EN 60079-14 covers the selection and installation of explosion-protected equipment, and BS EN 60079-17 covers the inspection and maintenance of these installations. There are also standards in the BS EN 500XX series which provide guidance on the construction requirements for electrical apparatus for potentially explosive atmospheres; for example, BS EN 50014 covers the general requirements and BS EN 50017 flameproof Ex 'd' enclosures. This topic is explained in more detail in Chapter 15.

Regulation 7 – Insulation, protection and placing of conductors

'All conductors in a system which may give rise to danger shall either:
(a) be suitably covered with insulating material and as necessary protected so as to prevent, so far as is reasonably practicable, danger; or

(b) have such precautions taken in respect of them (including, where appropriate, their being suitably placed) as will prevent, so far as is reasonably practicable, danger.'

The intention of Regulation 7 is to prevent electric shock and burn injuries from direct and indirect contact, and fire and explosion consequent on short circuits or leakage currents between circuit conductors or between circuit and other conductors.

Regulation 7(a) requires the insulation to be suitable, so it has to be chosen to suit the environmental conditions and the electrical stresses to which it will be subjected. In hot locations, for example, it has to be heat resistant, in damp places it has to be waterproof, and where there is the possibility of contamination by oils or other chemicals its insulating properties must not be adversely affected. In addition, it has to withstand the normal voltage and any transient high voltages. The 'as necessary protected' requirement is an additional safeguard, usually for protection against mechanical damage by, for example, running PVC-insulated wiring conductors in conduit or trunking or it could be for protection against other environmental hazards such as enclosure in a waterproof housing to prevent liquid contamination.

Regulation 7(b) refers to other precautions, including placing as an alternative to covering with insulation.

Regulation 8 – Earthing or other suitable precautions

'Precautions shall be taken, either by earthing or by other suitable means, to prevent danger arising when any conductor (other than a circuit conductor) which may reasonably foreseeably become charged as a result of either the use of a system, or a fault in a system, becomes so charged; and, for the purposes of securing compliance with this Regulation, a conductor shall be regarded as earthed when it is connected to the general mass of earth by conductors of sufficient strength and current-carrying capacity to discharge electrical energy to earth.'

As the public supply system employs neutral earthing, the most common method of compliance with this Regulation is to connect together any earth exposed and extraneous conductive parts by means of low impedance protective and bonding conductors. This is the EEBADS technique outlined in Chapter 3. The technique creates an equipotential zone so that when an earth fault occurs these conductive parts are raised to substantially the same potential with respect to the ground (earth). This condition persists until the protective device, be it a fuse or a circuit breaker, interrupts the circuit and clears the fault. Anyone in simultaneous contact with more than one of the conductive parts should not experience a shock because the parts are at about the same potential.

The Regulation requires the protective conductors to be effective under fault conditions. To ensure compliance, the prospective fault currents need to be ascertained and conductors selected which will carry these fault currents without damage until the protection operates to clear the fault.

If it is not possible to establish an equipotential zone because, for example, there is a conducting floor (such as a concrete floor) or because equipment is being used outdoors, supplementary or alternative measures must be taken. The use of residual current circuit breakers to provide sensitive earth leakage protection, in addition to the overcurrent protection, is one acceptable option; the RCD would detect earth fault currents, including shock currents flowing to earth, and rapidly interrupt the circuit. Indeed, it is general practice to ensure that socket outlets that will foreseeably be used to supply external equipment should have RCD-protection fitted.

As an alternative, precautions may be taken to prevent earth faults by segregating live conductors from exposed conductive parts by additional insulation. This creates a Class II system for the wiring and connected apparatus to prevent exposed conductive parts becoming charged.

For special applications, compliance may be achieved by other means. For example, the insertion of a 1:1 safety transformer in a mains voltage system would isolate the system on the secondary side from the earthed neutral system on the primary. Earthing one pole of the secondary side through an impedance so as to limit the potential earth fault current to no more than about 5 mA would create a system that would prevent electric shock injuries for a phase-to-earth fault. It is usual to employ a circuit that detects the flow of fault current and trips a circuit breaker controlling the supply. This type of system is often used in test areas.

Similar systems are used where continuity of supply is important and it is undesirable for earth faults to lead to the interruption of the supply. In these systems, the limited fault current flowing causes an audible or visual alarm rather than tripping a circuit breaker.

Fully separated systems, where a safety isolating transformer is inserted but the outgoing supply is not earthed, is another option frequently employed in production testing areas where an operator may have to handle or be exposed to contact with uninsulated live conductors. Measures must be taken to prevent one pole of the secondary side of the system becoming inadvertently connected to earth; the separated circuit should be kept as short as possible and the conductors should be well insulated and visible so far as possible. Periodic inspections and insulation resistance tests should be carried out to confirm the continuing earth-free integrity of the system.

Yet another alternative is the employment of a sufficiently low voltage to earth, suitable for the location's environment and usage, to avoid the possibility of a dangerous electric shock. The reduced low voltage system used on construction sites and described in Chapter 11 is an example.

Regulation 9 – Integrity of referenced conductors

'If a circuit conductor is connected to earth or to any reference point, nothing which might reasonably be expected to give rise to danger by breaking the electrical continuity or introducing high impedance shall be placed in that conductor unless suitable precautions are taken to prevent that danger.'

This Regulation aims to prevent open circuits or high impedances arising in the referenced conductors, which could cause hazardous potential differences between them and the reference point, and to prevent the flow of fault current in systems employing automatic disconnection protection techniques. In most cases, earth is the referencing point for supply systems but there are exceptions, such as motor vehicle wiring systems which are not earthed and which use the chassis as both a reference point and a common return. Other non-earthed systems connect the protective conductor to one of the supply poles.

As an example, in TN-C-S supplies to domestic premises a break in the combined neutral/earth (CNE) conductor of an overhead service line would cause the metalwork in the premises to become live at or about the supply voltage if any apparatus, such as a dishwasher, were to be connected and switched on. To prevent this type of hazard, the integrity of the CNE conductor, throughout the system, has to be maintained, so all joints have to be properly made and reliable, and fuses and solid state devices are prohibited in these conductors.

Regulation 10 – Connections

'Where necessary to prevent danger, every joint and connection in a system shall be mechanically and electrically suitable for use.'

Joints and connections are often the circuit locations most vulnerable to failure because they are frequently not entirely suitable for their purpose or because of poor workmanship or because of inadequate maintenance. An example may be the use of terminal blocks to joint cables – the blocks usually do not have strain relief and it is frequently the case that conductors are left exposed to be touched. This Regulation aims to ensure that joints and connections are both mechanically and electrically suitable.

In the main, only competent people should be used for making joints and connections in wiring systems. Skilled cable jointers, for example, should be employed to make joints in paper-insulated lead-sheathed steel wire armoured cables and the like. Trained people, such as electricians or non-electrically qualified but suitably instructed people, should be able to make satisfactory joints and connections in low voltage plastics-insulated wiring.

There is a range of factors that merit consideration when seeking to secure compliance with the Regulation. Some of these are:

- Environmental
    - □ *Vibration.* Consider the use of flexible multi-strand rather than single-wire conductors, and a soldered or compression joint rather than pinch screws that may shake loose. Plugs may need to be latched or otherwise secured into their sockets.
    - □ *Wet, corrosive or dusty conditions.* Seal or enclose the joint.
    - □ *Heat.* A cool wiring chamber may be advisable and the materials used should not be susceptible to the prevailing temperatures.
    - □ *Cold.* Anticondensation measures may be needed and sealants which harden and crack at low temperatures avoided.
    - □ *Dissimilar metals.* Joints and connections between dissimilar metals may require sealing to exclude air and moisture so as to avoid corrosion from electrolytic action.
- Mechanical
    - □ *Physical damage.* Protect joints from damage by position or enclosure. The clamping end of pinch screws should be designed to avoid damage to the conductor.
    - □ *Strain relief.* Secure conductors to relieve joints and terminations of strain.
    - □ *Connections.* Mechanical connections, in joints and terminals, should be designed to clamp conductors firmly and not slacken in service.
    - □ *Maintainability.* Connections that have to be periodically inspected and tested, such as earthing connections, should be readily accessible.
    - □ *Ventilation.* There should be adequate ventilation, or enclosures should be so designed that overheating does not occur.
- Electrical
    - □ *Insulation.* The insulation and air gaps must be suitable for the rated voltage and HV transients to prevent insulation failure and consequential short circuit between conductors.
    - □ *Cleanliness.* The conductors to be connected and the components of the joint must be clean. Plug pins and socket receptacles and the contacts of switches, circuit breakers and contactors must be clean and make effective contact.
    - □ *Contact resistance.* The contact area of the conductors in the connection must be sufficient to avoid overheating in service.

Regulation 11 – Means for protecting from excess of current

'Efficient means, suitably located, shall be provided for protecting from excess of current every part of a system as may be necessary to prevent danger.'

Excess current arises from overload, short circuit and earth faults. As the means of protection has to be efficient, it must respond sufficiently rapidly in interrupting the circuit so as to avoid danger. An overload trip, therefore, should be matched to the equipment it protects so that it will operate before dangerous overheating occurs. The device may also be the means of protection against short circuit so that it must be capable of safely interrupting a fault current that may be orders of magnitude greater than the overload current. This usually entails an inverse time characteristic, i.e. the greater the current the shorter is the tripping time, so as to limit the damage consequent on a fault and hopefully avoid a fire or explosion. The devices have an inherent current limiting characteristic that helps and some circuit breakers are designed with a higher impedance to exploit this feature.

Every part of a system has to be protected and the 'means' have to be suitably located. In practice, this means that there are a number of devices in series between the power source(s) and the loads. Those nearest the power source(s) usually need to be capable of interrupting higher fault currents than those further away. To ensure discrimination, time lag settings should provide for the device in the circuit nearest to the location of the excess current to operate before those nearer to the power source(s) so as to shut down only a part of the system. BS 7671 provides suitable guidance for locating the devices to protect every part of the system.

Regulation 12 – Means for cutting off the supply and for isolation

'(1)   Subject to paragraph (3), where necessary to prevent danger, suitable means (including, where appropriate, methods of identifying circuits) shall be available for:
(a)   cutting off the supply of electrical energy to any electrical equipment; and
(b)   the isolation of any electrical equipment.
(2)   In paragraph (1), 'isolation' means the disconnection and separation of the electrical equipment from every source of electrical energy in such a way that the disconnection and separation is secure.
(3)   Paragraph (1) shall not apply to electrical equipment which is itself a source of electrical energy but, in such a case as is necessary, precautions shall be taken to prevent, so far as is reasonably practicable, danger.'

The main purpose of Regulation 12 is that facilities must be provided for both switching off the power to circuits and for isolating circuits from their power sources. The Regulation also includes a requirement to identify circuits, which is usually done by applying labels to switchgear and control gear to identify the circuits that they are supplying. It is, perhaps, surprising how many circuits are not properly labelled, making it difficult for users and maintainers to know which switchgear controls which circuit, and creating the conditions for errors to be made when isolating circuits for work to be carried out on them.

The difference between the switching required by Regulation 12(1)a and the isolation required by Regulation 12(1)b needs to be understood. In both cases the supply has to be interrupted, but isolation has the additional requirement to secure the point of disconnection. This is usually done by locking a switch in the 'off' position, or by withdrawing a circuit breaker from the busbars, or by removing links or fuse links so that it is safe to work on the isolated part of the system. The means of isolation also serves to prevent inadvertent restoration of the supply while work is in progress. Push buttons controlling contactors meet the requirements for switching off but do not meet the requirement for isolation because there is a possibility of the contactor coil becoming energised and closing the contacts. Note that isolation normally requires a physical air gap to be present, so electronic circuits such as variable speed drives do not provide for isolation.

A question often arises about where isolators and disconnectors should be located and how many should be provided. There is no straightforward answer to this, because it largely depends on the circumstances. One school of thought is that each motor in a system should have its own local isolator positioned within a metre or so of the motor, ensuring that anybody working on a motor has direct control over its isolation. Whereas the provision of such local isolation has been standard practice for many years and will usually be the solution, others take the view that in some circumstances it is acceptable for 'group' isolation to be provided. For example, some manufacturers of integrated plant such as automated sawmills provide a single isolator to cover whole sections of the plant, each of which may contain many motors. Their logic is that if maintenance work has to be done on one motor in the section of the plant, the whole section has to be shut down, so one isolator will suffice. In this type of situation it is crucially important that the users of the plant ensure that maintenance personnel are properly trained on isolation procedures (see Regulation 13) and that they will use a group isolator when appropriate, despite the fact that it may be positioned tens of metres from the location of the work activity.

If there are means of both switching off and isolation, the isolator does not have to be capable of interrupting a fault or load current because the supply can be switched off before it is isolated, but if the isolator performs both functions it would have to be suitable for interrupting those currents safely. Some modern isolators have a set of auxiliary contacts that break before the main power contacts – the auxiliary contacts are used to signal the control system to open a power contactor or similar device so that the power is interrupted before the isolator contacts open.

The last part of the Regulation exempts power sources such as generators, primary and secondary batteries and charged capacitors which act as suppliers of energy. However, these power sources need to be protected to avoid danger.

Regulation 13 – Precautions for work on equipment made dead

'Adequate precautions shall be taken to prevent electrical equipment, which has been made dead in order to prevent danger while work is carried out on or near that equipment, from becoming charged during that work if danger may arise.'

Many accidents have happened, and continue to happen, when somebody working on or near a system that has been made dead experiences an electrical injury when the system is unintentionally switched on. This could be, for example, an electrician working on a system, or painters working in the vicinity of an overhead travelling crane's uninsulated power conductors. The aim of Regulation 13 is that measures should be taken to prevent this happening.

In some cases isolating a circuit using the facilities provided in accordance with Regulation 12 and applying a unique padlock and warning notices to the point of isolation will be an adequate precaution. However, more elaborate and considered precautions may be required where, for example, work is being carried out on dual circuit overhead line equipment. The circuit to be worked on may have been made dead but the other remains live; capacitive and inductive coupling between them could energise the isolated line, so temporary earthing would have to be applied to the isolated circuit.

Where it is foreseeable that more than one person may be working on plant or machinery that is isolated, the point of isolation needs to be controlled by all those working on the plant. A common solution is to use a multi-padlock hasp on to which a number of padlocks, typically up to eight, can be applied. This ensures that the isolator cannot be switched back on until all tradesmen have removed their padlock, signifying that all the work has been completed.

For work on most isolated HV equipment and for work involving, for example, the isolation of more than one supply source, it is advisable to employ a permit to work procedure as an additional safety measure. This is essentially a paperwork procedure in which a form is completed to identify the equipment on which work is to be done, how the equipment has been made safe, where the points of isolation and earthing are, and the limits of the work to be done. It is normally signed and issued by the person who made the system safe to a person who is going to carry out the work and who signs the form to indicate that he or she understands its contents. The permit to work form is cancelled once the work has been completed and the system restored to its operational configuration. The advantage of the procedure is that it imposes some structure and rigour on to isolation procedures on systems that have potential for serious injury. Given this, it is important for the procedure to be strictly adhered to and for its continuing implementation to be monitored and audited. Advice on this type of system of work is published by the HSE in Guidance Note HS(G)85 Electricity at work – Safe working practices.

Where a formal permit to work procedure is not used, the responsible person should carry out a risk assessment to devise a procedure that will ensure safety. In the case of the crane painters previously mentioned, for example, they should be told them that they should not start work in the vicinity of the bare conductors until the isolator has been locked off, the conductors have been tested to ensure that they are dead, and they have been given permission to proceed. When the work is finished the circuits should not be re-energised until the painters have been withdrawn from the work area.

Regulation 14 – Work on or near live conductors

'No person shall be engaged in any work activity on or so near any live conductor (other than one suitably covered with insulating material so as to prevent danger) that danger may arise unless –
(a) it is unreasonable in all the circumstances for it to be dead; and
(b) it is reasonable in all the circumstances for him to be at work on or near it while it is live; and
(c) suitable precautions (including where necessary the provision of suitable protective equipment) are taken to prevent injury.'

The main aim of Regulation 14 is to discourage dangerous live working, which over the years has been the cause of many serious and fatal accidents. The emphasis is on doing work on systems that are dead, with live work only being done if there is a strong case for it, and if the risks are acceptable, and if suitable precautions against injury are taken. Note the presence of the word 'and' between the subsidiary parts, meaning that all aspects must be considered before a decision is taken to authorise live work. Note also that the word 'work' is used; this has a very wide meaning and, contrary to popular belief, testing of live parts is a work activity that therefore falls within the scope of this Regulation.

Much live working is carried out in the electricity supply industry. This is mostly justified on the basis that it is frequently unreasonable to switch off the supply to many consumers or loads such as traffic lights and hospitals when work such as live cable jointing can be done safely by personnel who have the appropriate competencies and equipment. The enforcement authorities generally accept this argument but will invariably carry out searching enquiries when accidents happen during live working activity.

When an apparent need arises for work to be done on or near potentially dangerous uninsulated live conductors, the responsible person has to be able to justify it. This will involve carrying out a risk assessment that will include consideration of the consequences and the impact of switching off.

Where an electrical fault has halted a production process, for example, the maintenance electrician often needs to work on or near bare live conductors when using a test lamp or voltmeter to locate the fault. This type of live fault finding should only be done where it is not possible or practicable to trace the fault with the supply isolated. Where the work must be done live, the electrician must be competent in live fault finding activity. Test equipment with insulated and fused leads must be used, as well as tools that are insulated to avoid shorting out conductors at different potentials. Consideration should be given to shrouding off adjacent conductors at different potentials, including metalwork of panels and enclosures that may be at earth potential. Temporary barriers may need to be erected to minimise the chances of the electrician being disturbed or distracted. The presence of an accompanying person should also be considered – in the event of something going wrong, the second person's presence would ensure immediate help is available to switch off the supply, to drag the electrician clear, summon the rescue services, and perhaps render first-aid treatment.

Note that construction work adjacent to live overhead power lines comes within the remit of Regulation 14. The same considerations apply and the precautions against injury from contact with the lines are explained in Chapter 11.

HSE's previously-mentioned guidance note, HSG85, includes advice on the decision-making process as to whether work should be done live or dead.

Regulation 15 – Working space, access and lighting

'For the purposes of enabling injury to be prevented, adequate working space, adequate means of access, and adequate lighting shall be provided at all electrical equipment on which or near which work is being done in circumstances which may give rise to danger.'

Regulation 15 recognises that some electrical work is carried out in dangerous locations where injury should be prevented by, inter alia, having an adequate means of access and adequately and properly lit working space. The Regulation should be considered in conjunction with the 'suitable precautions' requirement of Regulation 14(c).

For compliance, it is advisable for the person responsible for the electrical wiring and apparatus such as switchgear and control gear to consider the requirements at the design stage of a project in conjunction with the architect or other space designer. The aim is to ensure that the requirements are met and thereby save what may be expensive modifications later. An example of the failure to do this arose a number of years ago when there was a vogue for flat-roofed building. The planners objected to the roof line being spoilt by lift houses and so the architects required the lift engineers to install their headgear in a small space on the top floor over the lift shaft between the top of the lift door and ceiling. In a number of cases, this resulted in unsafe working conditions for the maintenance staff and expensive alterations had to be made.

The space requirements will be minimised if equipment can be isolated for work, but even so where live testing is being carried out, or other forms of live work where live conductors are touchable, adequate space and lighting must be available for the work to be done safely. Some guidance on safe clearances may be gleaned from HSE's Memorandum of Guidance on the Regulations, HS(R)25, which reproduces the clearances in the former switchboard Regulation 17 of the Electricity (Factories Act) Special Regulations 1908 and 1944. The electricity supply industry companies' safety rules also contain information on safety clearances, usually as a function of voltage.

The space should also be adequate for safe movement in an emergency. For example, when horizontal drawout circuit breakers are withdrawn from a switchboard, there must be sufficient room for a person to escape from the switchroom in the event of a fire or explosion.

There has to be enough light for safe working. For most purposes, about 150 lux of general lighting should be adequate, supplemented if necessary by local lighting from handlamps or luminaires on pedestals. The lighting should be arranged so as to prevent dazzle and confusing shadows that may cause danger.

Regulation 16 – Persons to be competent to prevent danger and injury

'No person shall be engaged in any work activity where technical knowledge or experience is necessary to prevent danger or, where appropriate, injury, unless he possesses such knowledge or experience, or is under such degree of supervision as may be appropriate having regard to the nature of the work.'

Regulation 16 recognises that competence is an important prerequisite for 'any work activity' on electrical systems to be undertaken safely and that, where competence is not held, adequate supervision must be provided. The phrase 'any work activity' is obviously very broad so it would include those who plan the work and give instructions for its execution but do not themselves necessarily participate, as well as those who supervise and do the work on site. There are two areas of risk covered, i.e. the prevention of the risk of injury and the prevention of injury itself. Work on equipment that has been isolated from the supply and earthed is an example of the former because danger has been prevented, but live working entails some risk and is an example of the latter because the safety precautions counter the likelihood of injury occurring. Having said that, the difference between danger and injury in this context is a bit obtuse and the need for the distinction is not particularly obvious.

The matter of 'competence' is very important and frequently creates confusion and difficulty. For that reason, a chapter has been devoted to the topic, so the interested reader is directed to Chapter 14.

Part III: Mines

Since the Regulations in Part III only apply to mines and do not have general application beyond them, only a brief summary of the requirements of the Regulations is provided in the following paragraphs.

Regulation 17 – Provisions applying to mines only

Regulation 17 indicates that Regulations 18 to 28 and Schedule 1 apply to mines only. Schedule 1 deals with film lighting circuits. The intention is to apply a rigorous control over the temporary wiring and the apparatus used occasionally to provide illumination for filming below ground so as to prevent ignition of flammable dust or firedamp. In general, such apparatus and wiring are not explosion-protected nor as robust as equipment purpose-designed for mine use. The requirements include prenotification of the date for installation, inspection, test and supervision by a competent person and monitoring of the atmosphere for firedamp.

Regulation 18 – Introduction of electrical equipment

Regulation 18 applies where there is a possibility of flammable gas being present. The intention is to provide details of the electrical and ventilation proposals to the HSE Mines Inspectorate in sufficient time for it to check for safety before the work is done.

Regulation 19 – Restriction of equipment in certain zones below ground

Regulation 19 applies only to fiery mines. Its intention is to map the flammable zones, label them below ground to warn the occupants, and restrict electrical equipment to types that are approved or otherwise proved safe for use in flammable areas. However, portable equipment which does not comply, such as test instruments, may be used under Regulation 19(2)(g) provided its provisions are observed.

Regulation 20 – Cutting off electricity or making safe where firedamp is found either below ground or at the surface

The intention of Regulation 20 is to ensure that when a concentration of firedamp in excess of 1.25% by volume is detected, the electricity supply, at the location, is cut off to equipment which is not explosion-protected before the concentration reaches the lower explosive limit and becomes a hazard.

Regulation 21 – Approval of certain equipment for use in safety-lamp mines

Regulation 21 lists small items of apparatus required for the safety of persons underground, which are approved for taking below ground. The intention is to prevent other non-essential equipment from being taken underground in order to minimise the risk of ignition.

Regulation 22 – Means of cutting off electricity to circuits below ground

Compliance entails having a competent person available on the surface in touch with those underground so that he can switch off the supply, on request, in the event of danger. The switchgear has to be located on the surface and may be manually or remotely controlled. It should be capable of safely interrupting both its rated load current and the potential fault current arising from a short circuit fault.

Regulation 23 – Oil-filled equipment

Regulation 23 is intended to avoid the fire risk from ignition of the oil in oil-filled equipment. There is now an ample range of non-oil-filled equipment available to meet this requirement so new equipment should comply. Existing oil-filled apparatus may be retained until superseded, but should not be located in high risk locations such as in the vicinity of a coal face.

Regulation 24 – Records and information

In the mine office a schematic distribution diagram has to be displayed as well as a scale plan showing the location of all fixed equipment. In some cases it may be possible to combine the two plans but this should not be done at the expense of clarity. Where necessary, such as in substations and switchrooms, either the whole or a sufficient part of the schematic diagram has to be displayed to avoid danger, i.e. for the information of the electrical staff as an aid to safe operation. The diagrams and plans should be amended whenever alterations are made to the system.

Regulation 25 – Electric shock notices

The intention is to provide rescuers with resuscitation instructions and the procedure for summoning assistance. As these notices have to be posted at electrical equipment locations they can be conveniently and desirably extended to warn off would-be tamperers and also provide instructions on procedures in case of fire. Successful resuscitation is more likely if rescuers have been trained, so employees should be encouraged to learn the procedure.

Regulation 26 – Introduction of battery-powered locomotives and vehicles into safety-lamp mines

The locomotives and vehicles introduced into safety-lamp mines have to be of an approved type.

Regulation 27 – Storage, charging and transfer of electrical storage batteries

The requirements for storage, charging and transfer of batteries addressed by this Regulation relate mainly to traction-type batteries.

Regulation 28 – Disapplication of section 157 of the Mines and Quarries Act 1954

Regulation 28 abolishes the grounds for defence available under section 157 of the Mines and Quarries Act 1954 in respect of Regulations 18 to 27 and Schedule 1, and substitutes the 'due diligence' defence grounds of Regulation 29. It is applicable only for criminal proceedings.

Part IV: Miscellaneous and general

Regulation 29 – Defence

'In any proceedings for an offence consisting of a contravention of Regulations 4(4), 5, 8, 9, 10, 11, 12, 13, 14, 15, 16 or 25, it shall be a defence for any person to prove that he took all reasonable steps and exercised all due diligence to avoid the commission of that offence.'

Regulation 29 provides a necessary means of defence in criminal proceedings against alleged infringements of the listed Regulations. These are those that have absolute requirements and are not covered by the 'so far as is reasonably practicable' requirement. The intention is to provide some protection for persons who are guilty, for example, of a technical infringement, but who have done their best to comply and who are able to demonstrate this in court.

Regulation 30 – Exemption certificates

'(1)   Subject to paragraph (2) the Health and Safety Executive may, by a certificate in writing exempt -
(a)   any person;
(b)   any premises;
(c)   any electrical equipment;
(d)   any electrical system;
(e)   any electrical process;
(f)   any activity,
or any class of the above, from any requirement or prohibition imposed by these Regulations and any such exemption may be granted subject to conditions and to a limit of time and may be revoked by a certificate in writing at any time.
(2)   The Executive shall not grant any such exemption unless, having regard to the circumstances of the case, and in particular to –
(a)   the conditions, if any, which it proposes to attach to the exemption; and
(b)   any other requirements imposed by or under any enactment which apply to the case,
it is satisfied that the health and safety of persons who are likely to be affected by the exemption will not be prejudiced in consequence of it.'

The intention of Regulation 30 is to enable the responsible inspectorate, the HSE, to vary the legal requirements in particular circumstances where they are inappropriate. Exemptions are likely to be exceptional and conditional to ensure no deterioration in health and safety standards.

Regulation 31 – Extension outside Great Britain

'These Regulations shall apply to and in relation to premises and activities outside Great Britain to which sections 1 to 59 and 80 to 82 of the Health and Safety at Work etc. Act 1974 apply by virtue of Articles 6 and 7 of the Health and Safety at Work etc. Act 1974 (Application outside Great Britain) Order 1977 as they apply within Great Britain.'

Regulation 31 covers work over, on and under the territorial waters of Great Britain. It includes mines extending under the sea, the construction of oil and gas rigs and pipelines and their operation, diving, the loading and unloading, fuelling and provisioning of any vessel and any work on it except that done by the ship's company, and construction projects such as the Channel Tunnel and the Cardiff Bay barrage.

Regulation 32 – Disapplication of duties

'The duties imposed by these Regulations shall not extend to:
(a)   the master or crew of a sea-going ship or to the employer of such persons, in relation to the normal ship-board activities of a ship's crew under the direction of the master; or
(b)   any person, in relation to any aircraft or hovercraft which is moving under its own power.'

Regulation 32 is limited to the exclusion of the work on a seagoing ship by the master and crew and work on aircraft and hovercraft when moving under their own power.

Regulation 33 – Revocations and modifications

'(1)   The instruments specified in column 1 of Part 1 of Schedule 2 are revoked to the extent specified in the corresponding entry in column 3 of that Part.
(2)   The enactments and instruments specified in Part 11 of Schedule 2 shall be modified to the extent specified in that Part.
(4)   In the Mines and Quarries Act 1954, the Mines and Quarries (Tips) Act 1969(11) and the Mines Management Act 1971, and in regulations made under any of those Acts, or in health and safety regulations, any reference to any of those Acts shall be treated as including a reference to these Regulations.'

This Regulation refers to those parts of older law which are either revoked or modified.

COMMENT

The extension of specific electrical safety legislation beyond the boundaries of the factory gate has been the major contribution of these Regulations to the safety system in the UK. This, coupled with the Regulations' emphasis on working dead, and the specific provisions relating to maintenance, have had a very positive impact.

It has to be recognised that there is still widespread ignorance of the Regulations. Perhaps the situation would be improved if undergraduate electrical and electronic engineers at universities were to receive training and instruction on them. At least electricians who are undertaking apprenticeships in the electrical contracting industry now receive instruction on them so it can be anticipated that, as time passes, the electrical workforce will become increasingly aware of its legal responsibilities and the number of electrical injuries sustained at work will reduce even further.

Chapter 7

Law based on European Directives

INTRODUCTION

Legislation enacted in recent years to implement the provisions of European Directives has had a major impact in the health and safety field. Whilst only the Electrical Equipment (Safety) Regulations are targeted specifically and uniquely at electrical safety, many of them have requirements that have an influence in this field including, in this context, the safety of electrotechnical control systems on machinery. The purpose of this chapter is to identify those Regulations that have explicit and implicit electrical and control system requirements.

As observed in Chapter 4, some of the legislation stems from Directives made under Article 95 of the Treaty which established the European Community (Article 100a before the Amsterdam Treaty amended the Treaty of Rome). The main aim of this legislation is to ensure the free movement of goods throughout the European Union, but many of the Directives and associated UK legal provisions have a subsidiary but very important electrical safety content. Other legislation, derived from Article 137 Directives (previously Article 118a), is targeted at social provisions, including worker health and safety.

The application of European law in British law is a complex legal subject which attracts a considerable amount of debate among the legal profession; this debate is beyond the scope of this book. However, the following text describes the main UK legislation and is split into two blocks; the first block deals with legislation stemming from Article 95 and the second block deals with Article 137 legislation.

LEGISLATION STEMMING FROM ARTICLE 95 (100A) DIRECTIVES

As already noted, this legislation stems from the desire to establish a single market within the European Union (EU) inside the boundaries of which goods can be moved across national boundaries without undue hindrance. The philosophy behind the Directives is to establish a level playing field in which goods that are known to comply with appropriate European standards can be moved around on the basis that they are safe for use in any country within the EU. The CE mark, when attached to a product, is used to signify that the manufacturer of the particular product affirms that the product complies with all relevant Directives; in the main, there will also be paperwork in the form of Declarations or Certificates of Conformity on which the manufacturer details which Directives apply.

The Electrical Equipment (Safety) Regulations 1994

Scope

These Regulations were made on 15 December 1994 by the Secretary of State for Trade and Industry under the Consumer Protection Act 1987 to implement the amended Low Voltage Directive 93/68/EC of 22 July 1993. They extended and replaced the Low Voltage Electrical Equipment (Safety) Regulations 1989 and came into force on 9 January 1995, but were not applicable to equipment placed on the market before 1 January 1997. After this date all new equipment had to comply.

The Regulations apply to electrical equipment designed or adapted and supplied to operate between 50 V and 1000 V a.c. or between 75 V and 1500 V d.c. Note that the term 'supply' has a broad interpretation and would include, for example, electrical equipment that formed part of the rental agreement for rented domestic accommodation. The scope is the same as that covered by the Low Voltage Directive. The exclusions are equipment for use in explosive atmospheres, radiological and medical apparatus, parts for goods and passenger lifts, electricity meters, domestic plugs and sockets, electric fence controllers and equipment in aircraft, railways and ships.

The aim is to exclude unsafe electrical equipment from being marketed in the EU. To ensure this, the equipment has to comply with the relevant harmonised standard; in the absence of such a standard, to an International Electrotechnical Commission (IEC) standard; or, failing this, to a BSI standard. The manufacturer has to test the equipment himself to prove its safety or have it tested, preferably by an approved testing laboratory, and make a declaration of conformity with the relevant standards. He has to maintain technical documentation containing prescribed information and from 1 January 1997 had to apply the CE mark to the product, or its packaging, or its instruction sheet, or its guarantee certificate.

Compliance

Equipment made to a relevant standard or complying with other statutory safety requirements or certified by the manufacturer is regarded as sufficiently safe for compliance. If there are no applicable standards or statutory safety requirements the equipment is still acceptable if it is adequately insulated and/or earthed to prevent the risk of shock or fire. The Regulations apply to equipment as supplied and exclude any defects caused by improper installation, maintenance or use, but the maker must provide any necessary instructions.

Safety criteria

In Schedule 3 the safety criteria for compliance with Regulation 5 are listed and amount to:

(1) the provision of instructions by the manufacturer for the safe erection, connection, use and maintenance of the equipment;

(2) the equipment is to be designed so as to ensure persons and domestic animals are protected against electric shock and burn hazards and non-electrical dangers. Examples of the latter would include the provision of guards on chain saws, circular saws and grinders;

(3) that dangerous temperatures, arcs or radiation are not produced. On the face of it, this clause would prohibit electric are welding where dangerous electric arcs are necessarily produced for the process. The wording could, with advantage, be amended to make it clear that the production of such arcs under controlled conditions by instructed persons so as to minimise the hazard is acceptable;

(4) the equipment has to be robust enough to withstand rough handling and be suitable for use in the environmental conditions for which it is designed; and

(5) it has to be suitably protected against overload.

Guidance and enforcement

For domestic-type appliances, enforcement is by local authorities and usually by their trading standards officers. For equipment used by persons at work, enforcement is by HSE. The DTI has issued a guidance document, URN951626, to trading standards officers who enforce the Regulations in respect of domestic electrical apparatus. Complaints about commercial and industrial products are handled by the HSE.

The penalties for infringement are a maximum fine at level five on the standard scale, or imprisonment, not exceeding six months where there is a risk of death or injury to a person, or three months otherwise, or both a fine and imprisonment.

The Supply of Machinery (Safety) Regulations 1992

Scope

The Supply of Machinery (Safety) Regulations 1992 were made by the Secretary of State of the Department of Trade and Industry to implement the Machinery Directive 89/392/EC as amended by Directive 91/368/EC and the 'CE Marking Directive' (93/68/EC). The Regulations were amended by the Supply of Machinery (Safety) (Amendment) Regulations 1994 which widened the scope to include machinery for lifting persons and safety components for machinery. The 1992 Regulations came into force on 1 January 1993 and the Amendment Regulations on 1 January 1995. There was, however, a transitional period for the 1992 Regulations up to 1 January 1995 to enable manufacturers to adapt to the new Regulations provided the machinery complied with UK safety legislation in force on 31 December 1992. For items covered by the Amendment, the transitional period extended to 1 January 1997 provided they complied with the UK safety legislation in force on 14 June 1993.

The Regulations are concerned with the safety of machinery and apply to all machinery placed on the market in Europe, including those imported from non-EU countries. They cover all the hazards associated with the installation, use, cleaning, maintenance and dismantling of machinery; this includes electrical hazards.

There is a long list of exclusions. These include manually powered machines except those used in lifting and lowering loads; passenger and goods transporters used on public roads, rail and in the air or on water; passenger and goods lifts; and other machinery covered by other Regulations such as electrical equipment subject to the Electrical Equipment (Safety) Regulations 1994 where the risks are predominantly electrical. The Regulations are not, therefore, predominantly concerned with electrical safety, but where machines are electrically powered or have electrotechnical control systems, the integrity and safety of these systems need consideration.

Compliance

The Regulations have requirements relating to the application of a CE mark, the generation and retention of a technical construction file containing prescribed information, and the drawing up of declarations of conformity or declarations of incorporation (depending on whether the product is a complete machine that can operate on its own or is destined for incorporation into other machinery). They also detail the acceptable conformity assessment procedures which may include, for certain types of more dangerous machinery listed in Schedule 4 to the Regulations, submitting the product or its technical file to an approved body.

The Regulations have attached to them an extensive list of requirements called Essential Health and Safety Requirements (EHSRs) which must be complied with by manufacturers or suppliers. Among them are quite generic safety requirements covering electrical systems, including:

(1) protection against hazards from the electricity supply and static electricity;
(2) the provision of means for isolation from energy sources;
(3) safety and reliability of control systems;
(4) stopping devices;
(5) failure of the power supply

One route to compliance with these EHSRs is designing and building machines, devices or components in conformity with the provisions of a harmonised standard describing the particular machine, device or component. The standards are produced by Technical Committees and their working groups of experts within CEN, the European standards making body. The working groups are formed from experts from nations within the EU and will generally represent manufacturers and users of machines, and the regulatory bodies such as HSE. There are three generic types of harmonised standards: A, B and C.

(1) Type-A standards are basic safety standards that have broad application to all machinery and safety components. Perhaps the most important of these is EN 292 Safety of machinery – Basic concepts, general principles for design, Parts 1 and 2.
(2) Type-B standards are subdivided into two subtypes: Type-B1 and Type-B2. Type-B1 standards are application standards, an example being EN 60204-1 Safety of machinery, Electrical equipment of machines, General requirements. This standard provides guidance on the means of providing protection against electrical injury on machinery as well as on the provision of safe control systems; it can therefore be considered as a basic standard covering electrical safety principles. Type-B2 standards cover particular safety components and devices; an example is EN 61496-1 Safety of machinery, Electrosensitive protective equipment, General requirements.

(3) Type-C standards relate to specific types or groups of machines. Manufacturers who build machinery in conformity with a Type-C standard have a 'presumption of conformity' with the EHSRs.

The EN standards, once published, are named in the Official Journal of the European Commission. Once this happens, they can be published as national standards – in the case of the UK, BSI will publish them and market them as BS ENs.

Manufacturers will clearly be keen to have access to Type-C standards for their particular type of machinery. Unfortunately, progress with the production of Type-C standards, in particular, has been quite slow and this has made it difficult for manufacturers who are aiming to comply with the EHSRs and the Regulations. Most of the important Type-A and Type-B standards have been produced and published – indeed, some of them have either been updated or are about to be.

Guidance

The DTI has produced a number of guidance documents on the Regulations, which can be obtained directly from the DTI itself or through HMSO bookshops.

Enforcement

Enforcement is by the HSE for machinery and safety components for use at work and by trading standards officers for items for domestic use.

The Electromagnetic Compatibility Regulations 1992

Scope

These Regulations were made by the Secretary of State of the Department of Trade and Industry to implement the Electromagnetic Compatibility Directive (89/336/EC), coming into force on 28 October 1992. There was a transitional period from then until 31 December 1995, during which manufacturers of affected equipment had the option of complying either with the Regulations or with legislation in force on 30 June 1992 in the member state in which the product was marketed. They were updated by the Electromagnetic Compatibility (Amendment) Regulations 1994 to enact the provisions of the 'CE Marking Directive' (93/68/EEC). A revised EMC Directive is expected to be promulgated by the European Council in the first half of 2002, with the intention of clarifying some aspects of the original Directive.

The Regulations apply to electrical and electronic apparatus which has the potential to emit electromagnetic radiation which may cause harmful interference with other electrical or electronic apparatus, or which may be affected by radiation emitted by other apparatus. The Regulations refer to 'Electromagnetic disturbance', meaning any electromagnetic phenomenon which is liable to degrade the performance of relevant apparatus. Quoted examples of phenomena are:

■ electromagnetic noise;
■ unwanted signals; and
■ changes in the propagation medium.

In reality, these will include conducted phenomena such as harmonics, signalling voltages, oscillatory transients and voltage fluctuations; and radiated phenomena in the electromagnetic spectrum.

A signal or emission which is a necessary function, or consequence of the operation, of relevant apparatus is not taken to be electromagnetic disturbance if, in relation to that apparatus, the signal or emission is permitted and does not exceed specified limits.

Exclusions

There is a very long list of exclusions covering, among other things:

(1) Apparatus for export to a third country outside the EU;
(2) Installations where two or more combined items of relevant apparatus or systems are put together at a given place (whether or not in combination with any other item) to fulfil a specific objective but not designed by the manufacturer (or manufacturers, where the items are made by different manufacturers) for supply as a single functional unit;
(3) spare parts;
(4) second-hand apparatus, as defined in the Regulations;
(5) electromagnetically benign apparatus;
(6) apparatus for use in a sealed electromagnetic environment;
(7) radio amateur apparatus which is not available commercially;
(8) military equipment;
(9) active implantable medical devices
(10) medical devices;
(11) electrical energy meters;
(12) spark-ignition engines of vehicles;
(13) non-automatic weighing instruments;
(14) telecommunications terminal equipment.

Requirements

The general requirements are that relevant apparatus has to be so constructed that:

- the electromagnetic disturbance it generates does not exceed a level allowing other relevant apparatus to operate as intended; and
- it has a level of intrinsic immunity which is adequate to enable it to operate as intended, when it is:
  - □ properly installed and maintained; and
  - □ used for the purpose for which it was intended.

In addition, the apparatus must bear the CE mark and have a certificate of conformity.

Compliance

Compliance with the Regulations can be secured through one of three routes:

(1) The manufacturer may self-certify that the product complies with an appropriate European harmonised standard. There is a significant number of standards that can be used; the main generic ones are:
  - BS EN 50081-1 : 1992 Electromagnetic compatibility. Generic emission standard. Residential, commercial and light industry.
  - BS EN 50081-2 : 1994 Electromagnetic compatibility. Generic emission standard. Industrial environment.
  - BS EN 50082-1 : 1998 Electromagnetic compatibility. Generic immunity standard. Residential, commercial and light industry.
  - BS EN 50082-2 : 1995 Electromagnetic compatibility. Generic immunity standard. Industrial environment.

  In addition to these generic standards, there are product specific standards which specify EMC requirements.
(2) If there are no relevant standards, or the manufacturer chooses not to use the standards route, then a technical construction file approved by a competent body can be produced.
(3) The manufacturer can arrange for a notified body to produce and issue an EC type examination certificate. This route is compulsory in the case of radio equipment.

Enforcement

Enforcement responsibility rests primarily with trading standards departments of local authorities for most industrial and consumer products.

Trading standards officers will generally only react to complaints or incidents, rather than taking proactive action in respect of the Regulations.

The DTI enforces the Regulations in respect of radio-communications transmitting equipment and the Civil Aviation Authority has responsibility for specific wireless telegraphy apparatus.

The Equipment and Protective Systems Intended for Use in Potentially Explosive Atmospheres Regulations 1996

Scope

The Equipment and Protective Systems Intended for Use in Potentially Explosive Atmospheres Regulations 1996 were made by the Secretary of State for Trade and Industry, coming into force on 1 March 1996. They implement the provisions of Directive 94/9/EC, the so-called ATEX Directive, and they will be referred to here as the ATEX Regulations.

The Regulations apply to manufacturers and suppliers of equipment that is intended for use in potentially explosive atmospheres. A very important consideration is that the term 'equipment' extends beyond the electrical equipment covered in existing law on this topic. 'Equipment' is defined as machines, apparatus, fixed or mobile devices, control components and instrumentation thereof and prevention or systems which, separately or jointly, are intended for the generation, transfer, storage, measurement, control and conversion of energy or the processing of material and which are capable of causing an explosion through their own potential sources of ignition. The Regulations cover such equipment used below ground, on the surface and on fixed offshore installations. This means, for example, that an internal combustion engine falls within the remit of the Regulations as much as an electrical switch or luminaire.

Another important change on 'traditional' legislation is that the ATEX Regulations extend to all types of potentially explosive atmospheres, including those consisting of combustible dust. Indeed, an explosive atmosphere is defined as mixtures with air, under atmospheric conditions, of flammable substances in the form of gases, vapours, mists or dusts in which, after ignition has occurred, combustion spreads to the entire unburned mixture, i.e. there is an explosion. This definition has the effect of excluding explosions as a result of increases in pressure, or as a result of chemical interactions.

The main intent of the legislation is to remove the need for documentation and testing for each individual European market. Manufacturers only have to CE-mark their products once to show compliance with the Directive, with the product then being able to be traded freely throughout the EU. However, there are obvious benefits from the safety perspective.

There is a transitional period leading up to 1 July 2003 during which manufacturers can choose to satisfy the provisions of the new Regulations, or comply with the existing Explosive Atmospheres Directives, or comply with the general provisions of the HSW Act, section 6.

The DTI has proposed an amendment to the Regulations. The main reason for the amendment is to include within the scope of the Regulations the taking into service of equipment as well as the placing on the market, so as to cover manufacture of products by the end user. In addition, the amendment addresses products which are manufactured before July 2003 that are stockpiled by distributors and end users for sale or use from July 2003; in particular, where the products require assembly, modification, bespoke installation or modification, or where the product remains in the supply chain prior to July 2003.

Exclusions

Some types of equipment are excluded from the scope of the Regulations. To all intents and purposes this relates to domestic and non-commercial equipment, so confining the Regulations to equipment intended for use at work. Equipment and protective systems where the explosive hazard is entirely due to the presence of explosive or unstable chemical substances are also excluded.

Duties

The main duty on manufacturers or suppliers is to ensure that equipment destined for use in potentially explosive atmospheres satisfies relevant Essential Health and Safety Requirements, which are listed in Schedule 3 of the Regulations. The Requirements relate to three groups: common requirements, supplementary requirements for equipment, and supplementary requirements for protective systems. Protective systems are defined as design units which are intended to halve incipient explosions immediately and/or to limit the effective range of explosion flames and explosion pressures. Protective systems may be integrated into equipment or separately placed on the market for use as autonomous systems.

There are additional duties relating to the information and instructions to be provided, as well as the markings that must be applied to the equipment. All equipment and protective systems must be CE-marked once their conformity to the EHSRs has been assessed.

Categorisation

The Regulations define two main groups for equipment, following the traditional route: Group I equipment for use in underground mines and other high hazard areas where very high levels of protection would be justified; and Group II equipment intended for use in areas where explosive atmospheres from gases, vapours, mists and/or combustible dusts could exist, for varying lengths of time. The groups are then subdivided into a range of different categories. For Group II equipment, there are three categories (1, 2 and 3) which are broadly equivalent to the well-known Zones 0, 1 and 2; see Chapter 15 for information on zoning.

Conformity

The Regulations provide a range of means for assessing conformity. However, as far as electrical equipment is concerned, conformity assessment through a notified body will be the norm, following much the same EC type examination processes and procedures that have been used in the past. There are, however, additional requirements for product quality control and third party audits. In the UK, the Electrical Equipment Certification Service (EECS) and SIRA Test and Certification Ltd already provide conformity assessment services, basing their assessments on the harmonised standards that already exist for explosion-protected electrical equipment; the main standards are identified in Chapter 15.

The situation is less clear for non-electrical equipment because there are insufficient harmonised standards to cover the full range of equipment.

Enforcement

HSE is the sole enforcing authority for the Regulations in the UK.

LEGISLATION STEMMING FROM ARTICLE 137 (118A) DIRECTIVES

As already noted, this legislation stems from the social policy elements of the Treaty, with particular emphasis on worker health and safety. The following text covers the main legislation falling within this category.

The Management of Health and Safety at Work Regulations 1999

These Regulations are intended to implement the Framework Directive (89/391/EC), although they also enact Directives relating to pregnant workers, young people and temporary workers at work.

Although these Regulations have wide-ranging provisions relating to the management of health and safety (including specific requirements relating to young persons and new and expectant mothers), the most important provision in the context of this book is Regulation 3. This covers the duties that employers and the self-employed have for carrying out 'suitable and sufficient' risk assessments. To identify the measures he needs to take to comply with the requirements and prohibitions imposed upon him by or under the relevant statutory provisions, an employer must assess:

■ the risks to the health and safety of his employees to which they are exposed while they are at work; and
■ the risks to the health and safety of persons not in his employment arising out of or in connection with the conduct by him of his undertaking.

In the context of this book, this would include work activities such as live working and the isolation of high voltage systems.

A self-employed person must assess:

■ the risks to his own health and safety to which he is exposed while he is at work; and
■ the risks to the health and safety of persons not in his employment arising out of or in connection with the conduct by him of his undertaking, for the purpose of identifying the measures he needs to take to comply with the requirements and prohibitions imposed on him by or under the relevant statutory provisions.

Where an employer has five or more employees, he must record:

■ the significant findings of the assessment; and
■ any group of his employees identified by it as being especially at risk.

So, whereas none of the specific electrical safety regulations make any mention of risk assessment, there is a duty under these Regulations to conduct such assessments and record them if five or more people are employed. As previously observed, it is difficult to see how duty holders can make judgements about the reasonable practicability of measures without carrying out risk assessments. A good source of advice on carrying out risk assessments is the HSC/HSE publication *5 Steps to Risk Assessment*.

The Regulations are enforced by HSE inspectors and local authority environmental health officers.

Provision and Use of Work Equipment Regulations (PUWER) 1998

These Regulations, which came into force on 5 December 1998, replaced the original Provision and Use of Work Equipment Regulations (PUWER) 1992. They enact the Work Equipment Directive (89/655/EC) as amended by the Amending Directive to the Use of Work Equipment Directive, which extended the original Directive to include mobile work equipment, lifting equipment (although lifting equipment is not actually covered in PUWER 98) and the inspection of work equipment. They have the main objective of ensuring that work equipment should not result in risks to health and safety. The term 'work equipment' has very broad meaning, and includes all electrical and electronic apparatus and equipment used at work. The Regulations are supported by an Approved Code of Practice and guidance material published in HSC/HSE document L22.

As far as electrical and electronic systems are concerned, there are a number of relevant Regulations that are of interest.

Regulation 3, on inspection, introduces a requirement for work equipment to be inspected when it is first put into use and then at routine intervals thereafter if the equipment is likely to deteriorate in its conditions of use. The ACOP specifies that the inspections should be done by a competent person, without giving very much guidance on the meaning of the term 'competent person'; however, see Chapter 14 for information on competence. The Regulation also requires the inspection to be recorded; although the recording method is not specified, or what the contents of the records should be, there is information on this in the guidance material.

There is frequently a question raised when referring to inspection of electrical apparatus and equipment about the extent to which tests should be carried out. The stated intention of the Regulation is to ensure that the equipment is safe to operate, in which case it would be reasonable to argue that properties such as insulation resistance and earth continuity should be tested. However, it would also be reasonable for the competent person carrying out the inspection to determine what maintenance has been done on the equipment to comply with the Electricity at Work Regulations, Regulation 4(2). If he is able to confirm that appropriate inspections and tests of the precautions taken against electrical injury have been carried out to a satisfactory standard, he may be able to dispense with this type of test on the basis that they have been carried out by the user. In that case, he would restrict himself to a visual examination of the insulation, terminal connections and so on. He would, however, be well advised to carry out such tests himself if there is any doubt about the integrity of the electrical insulation and earthing.

Functional tests of safety-related control systems, such as emergency stops and guard interlocks, should always be carried out as part of an inspection carried out under this Regulation.

Regulations 14 to 18 have particular requirements relating to control systems, including those that carry out safety functions. This topic is covered in detail in Chapter 13 but the following summarises the requirements.

Regulation 14 covers the requirement for equipment to have some form of start control to ensure that it can only be started as a result of a deliberate action. It would not be safe, for example, for a hand-held electric drill to start as soon as the power to the drill is switched on – there must be a start button to ensure that it will only start when the user wants it to start. A similar argument holds true when the machine can change its operating characteristics, such as speed, when that might lead to a dangerous condition for the operator. An example of this would a robotic palletising machine that moves at crawl speed when it is in 'teaching' mode – it would not be safe if the machine were to suddenly move at full operational speed unless and until the operator has given a specific command to the machine's control system to enter 'operational' mode. It is accepted, however, in Regulation 14(3), that many automated machines will start, stop and change mode as part of their operating cycles – this is acceptable so long as the machine is adequately safeguarded to prevent injury to the operators.

Regulation 15 stipulates that equipment must be provided with some means to bring it to a safe condition in a safe manner. This will normally be some form of stop control that, when activated, brings those parts that have dangerous motion or other dangerous feature such as high temperature or pressure to a safe state. This does not need to be an instantaneous stop – there are many examples of machines in which the control system will need to go through a sequence of control actions to bring the equipment to a safe condition in a controlled manner. The technical requirements for stop devices and functions can be found in BS EN 60204-1, in which the stop function is categorised in one of three categories: 0, 1 and 2. A category 0 stop is one in which power is removed immediately, usually by electromechanical means such as a contactor; a category 1 stop is one in which the machine is brought to a safe stop under control means and then power is removed, usually by electromechanical means; a category 2 stop is one in which the machine is brought to a controlled stop and power is retained on the system once the safe condition has been achieved.

Regulation 16 requires that work equipment must be provided with one or more emergency stop controls, but only where they can and will contribute to risk reduction. The decision on whether or not to provide emergency stop devices and, if so, how many and where they should be located should form part of the risk assessment carried out on the machine. The requirements for emergency stop actuators are published in BS EN 418. Note that category 2 stops must not be used for the emergency stop function, although either category 0 or 1 stops may be so used.

Regulation 17 covers the suitability of control devices on work equipment. This addresses the positioning, visibility, identification and other characteristics of control buttons, switches and other actuators.

Regulation 18 covers the safety integrity of control systems that carry out safety functions. Such control systems need to have sufficient integrity, in terms of fault tolerance and reliability, when compared to the amount of risk reduction they are aiming to achieve. This subject is treated in detail in Chapter 13.

Regulation 19 covers the means for isolating sources of energy and, as far as electrical energy is concerned, has the same effect as the Electricity at Work Regulations, Regulations 12 and 13.

The Regulations are enforced by HSE inspectors and local authority environmental health officers.

The Personal Protective Equipment at Work (PPE) Regulations 1992

The relevant Directive which these Regulations implement is 891656/EC. They were made on 25 November 1992 and came into force on 1 January 1993 and are administered by the HSE and local authorities.

The onus is on the employer or self-employed to assess the risks in the workplace and to provide and maintain appropriate personal protective equipment which is worn by the worker. In the electrical risk context this would comprise, for example, insulating boots and gloves for live working, and semiconducting footwear and clothing where it is necessary to avoid static discharges, such as in operating theatres or during the electrostatic spraying of flammable paints and powders. For arc welding, gloves, aprons and footwear to resist hot metal spatter are required and a helmet with a visor and suitable lenses to protect the face, reduce the glare and cut out the UV radiation. Where noxious fumes are produced, and/or in confined conductive spaces, the helmet may need connecting to a fresh air supply and the operator may need to be provided with insulating clothing and footwear to counter the electric shock hazard.

The employer has to inform employees of the hazards the equipment is designed to avoid, train them in its use, and ensure that is properly used. More guidance is published in HSE guidance booklet L25.

The Construction (Design and Management) Regulations 1994

Made on 19 December 1994 by the Secretary of State of the Department of Employment to implement Directive 92157/EC, the Construction (Design and Management) Regulations 1994, commonly known as the 'CDM Regulations' came into force on 31 March 1995. They detail and extend the requirements of the Health and Safety at Work Act by introducing a mandatory safety procedure for construction work which, it is hoped, will counter the somewhat cavalier attitude to safety that has been endemic in the construction industry and will result in an improvement in its appalling accident record.

The Regulations place duties on clients, principal contractors, contractors, designers and planning supervisors relating to the health and safety of construction activity. In general terms, the Regulations apply to construction work (including demolition and dismantling) in non-domestic premises which will last longer than 30 days, will involve more than 500 person days of construction, and in which the number of people involved in the construction work is five or more.

The client's duties include selecting and appointing a competent planning supervisor and principal contractor; satisfying himself that designers and other contractors are also competent; ensuring that the planning supervisor is provided with relevant health and safety information; ensuring that work on site does not begin until the principal contractor has prepared a satisfactory health and safety plan; and ensuring that the health and safety file is available for inspection on completion of the project.

The planning supervisor's main responsibilities include ensuring that the design considerations take full account of health and safety considerations; ensuring designers cooperate with regard to health and safety arrangements; giving advice on health and safety matters to clients and designers; and ensuring that a health and safety file is prepared and kept up to date and delivered to the client at the end of the project.

Principal contractors must do all they can to ensure that contractors and employees comply with the health and safety plan; that all the contractors on the project cooperate with each other; and that only authorised persons work on the site. They must also keep the planning supervisor up-dated and ensure that people are informed about health and safety rules.

Designers, such as architects, must ensure that their designs minimise the risks to persons carrying out the construction or cleaning work, and to persons who may be affected by the work.

So far as the electrical subcontractor is concerned, if the installation is designed to comply with the requirements of the Electricity at Work Regulations 1989 and in such a manner as to minimise the risks entailed in its construction and subsequent maintenance, it will be likely to be in compliance with the CDM Regulations. It is important, however, that the contractor cooperates fully with the other CDM duty holders and that he employs staff who are competent both in terms of the Electricity at Work

Regulations and the CDM Regulations. In this respect he has to cooperate with other designers and the planning supervisor. One obvious requirement is to ensure safe access to those parts of the installation that require periodic attention. In his role as a subcontractor, he has to provide information for incorporation in the health and safety plan and health and safety file, cooperate with the principal contractor and ensure that his employees are properly briefed on health and safety relevant to the site. Where the electrical contractor is the only contractor, such as may occur for a rewiring project, he becomes the principal contractor and has to shoulder the relevant responsibilities.

Enforcement is by the HSE, who have published an approved code of practice.

There is no doubt that, since their inception, the Regulations have had a major impact in the management of construction activity and the way in which health and safety arrangements are planned and recorded. Some people argue that they have increased the amount of paperwork and bureaucracy involved in construction projects. However, it is a mute point whether or not these changes have worked their way down to the 'coal face' and had an impact on the way in which the tradesmen themselves conduct their activities and take account of health and safety arrangements. Certainly, in 2001 there appears to have been no reduction in the number of construction-related deaths and serious injuries.

The Construction (Health, Safety and Welfare) Regulations 1996

The Construction (Health, Safety and Welfare) Regulations 1996 were made on 14 June 1990 and came into force on 2 September 1996 to implement part of Directive 92157/EC and should be read in conjunction with the Construction (Design and Management) Regulations. There are no specific electrical requirements, but Regulation 17 states that plant and equipment has to be safe and maintained in a safe condition. This would include the electrical installation. Standards for electrical installations on construction sites are explained in Chapter 11.

Protection of workers potentially at risk from explosive atmospheres Directive

Directive 1999/92/EC is an Article 137 Directive laying out the minimum requirements for improving the safety and health protection of workers potentially at risk from explosive atmospheres. As such it complements the Article 95 Directive evolved into The Equipment and Protective Systems Intended for Use in Potentially Explosive Atmospheres Regulations 1996. The Directive was adopted by the Council of Ministers on 6 December 1999.

The Directive places duties on employers, in decreasing order of preference, to take measures to protect the health and safety of their employees by preventing the formation of explosive atmospheres, by avoiding the ignition of explosive atmospheres, and by mitigating the effects of an explosion. They must also take measures against the propagation of explosions. In order to achieve this, the employer must assess the risks and classify the workplace into zones (0, 1 and 2), the nature of which follow the well-known zonal classification scheme explained in Chapter 15. The Directive also describes three zones (20, 21 and 22) relating to the presence of combustible dusts. This is the first time that a specific duty to carry out a zone classification exercise will appear in a statutory instrument.

The employer must draw up an Explosion Protection Document to demonstrate that the risks have been assessed and to explain the precautions to be taken to control the risks. He must also coordinate the implementation of health and safety measures with other employers.

The Directive relates to 'work equipment' which, as for The Equipment and Protective Systems Intended for Use in Potentially Explosive Atmospheres Regulations 1996, is not restricted to electrical equipment.

The Directive is meant to be enacted by 30 June 2003, but there has been a debate within the UK about how it should be enacted, especially in relation to the common requirements between it and the Chemical Agents Directive. It is anticipated that the Regulations enacting the Directive will appear during 2002.

Chapter 8
Electricity Supply Legislation

INTRODUCTION

The electrical hazards of shock, burn, fire and explosion were appreciated as far back as the 1880s when electricity was first used as a source of energy. The Electric Lighting Act 1882 empowered the then Board of Trade to impose safety requirements 'for securing the safety of the public from personal injury or from fire or otherwise'. The Board of Trade, their successors the Electricity Commissioners, the Department of Energy and now the Department of Trade and Industry, have all used the power to issue safety regulations, including the Electricity Supply Regulations 1988. Some of this legislation was still extant, some had been amended and some repealed up to the inception of the Electricity Act 1989 where, in Schedule 18, the opportunity was taken to repeal it, but the transitory provisions of Schedule 17 preserved the Electricity Supply Regulations 1988.

In February 2001 the DTI issued a consultation paper for the promulgation of a set of new Regulations to replace the Electricity Supply Regulations 1988, to be known as The Electricity Safety, Quality and Continuity Regulations. The main driver behind the new Regulations is the break up of the electricity supply industry since privatisation and the separation of businesses under the Utilities Act 2000 into those concerned with electricity distribution and those concerned with supplying electricity to consumers. The commercial competition stemming from these changes has had a major impact on the structure of the industry and the way in which it operates, so much so that the 1988 Regulations were perceived to be increasingly irrelevant, with a consequential need for them to be revised to ensure continuing protection of the public and continuity and quality of supply.

The publicised intention had been for the new Regulations to be promulgated in October 2001 but this timescale was not achieved, presumably because the consultation exercise produced so many comments that they could not be resolved in time. A revised set of proposals was issued for limited consultation in October 2001, with the stated aim of promulgating the new Regulations in June 2002.

This chapter describes the provisions of the 1988 Regulations and the main requirements of the new Electricity Safety, Quality and Continuity Regulations, highlighting the principal changes. Since the new regulations had yet to be enacted when this chapter was written, their description is based on the contents of the DTI consultation documents; the final published regulations may differ slightly from those described, although any changes are likely to be minor.

Since the electricity supply legislation is raised by the Department of Trade and Industry, the Secretary of State for Trade and Industry has the enforcement responsibilities, and these are exercised by the DTI's Engineering Inspectorate. In securing public safety, it is mainly concerned with the overhead transmission and distribution systems, the overhead service lines to consumers' premises and the apparatus up to the consumers' terminals, as all of these are accessible to the public. There is some overlap of responsibility with the HSE whose concern, under the Health and Safety at Work Act and other health and safety legislation, is the safety of workers and of the public arising from work activities in the supply industries.

THE ELECTRICITY SUPPLY REGULATIONS 1988

These regulations are applicable to both public and private electricity suppliers and were a revision of the Electricity Supply Regulations 1937 'for securing the safety of the public and for insuring a proper and sufficient supply of electrical energy' and of the Electricity (Overhead Lines) Regulations 1970. They were made under section 16 of the Energy Act 1983, with a commencement date of 1 October 1988, but existing works which did not comply were exempt from some of the Regulations until such time as material alterations were made to them. Consequent on the Electricity Act 1989, minor amendments to the Regulations were made under the following legislation (Statutory Instruments):

■ The Electricity Supply (Amendment) Regulations 1990 – (SI) 1990/390
■ The Electricity Supply (Amendment) Regulations 1992 – SI 1992/2961
■ The Electricity Supply (Amendment) Regulations 1994 – SI 1994/533
■ The Electricity Supply (Amendment) (No 2) Regulations 1994 – SI 1994/3021
■ The Electricity Supply (Amendment) Regulations 1998 – SI 1998/2971

The Regulations are intended to ensure the reliability and quality of the electricity supply and to safeguard the public, livestock and domestic animals against the hazards of burn, shock, injury from mechanical movement and fire from the generation, transformation and supply or use of electricity.

Part I – Introductory

Part I of the Regulations is mainly given to providing definitions, or interpretations, of terms used in the rest of the Regulations. It is worth observing that throughout the Regulations there are references to phase and neutral but not to positive and negative, and although d.c. is not specifically excluded, the only reference to it, apart from that in the Interpretations, is in Regulation 30(6) which requires the supplier not to alter the polarity of a d.c. supply without the agreement of the consumer.

Part II – Connection with earth

Regulations 4 to 8 refer to the continuity of the supply neutral conductor, general requirements for connection with earth, multiple earthing of the neutral, protective multiple earthing (PME) supplies on consumers' premises, and metalwork earthing. The main duties are to:

■ Take all reasonable precautions to ensure continuity of the supply neutral conductor.
■ Connect electricity supply systems to earth as close as possible to the source of voltage and meet other general earthing requirements set out in Regulation 5.
■ Ensure that where the neutral conductor of a distributing main is connected to earth at multiple locations, the minimum restrictions on the cross sectional area of the neutral conductor set out in Regulation 6 are satisfied.
■ Where a combined neutral and earth conductor is used to supply a consumer's premises, the supplier must meet the protective multiple earthing requirements of Regulation 7, as well as ensure that the consumer's installation meets the criteria set out in the same Regulation for equipotential bonding and earthing. The neutral conductor of PME supplies must not be connected through to metalwork in caravans or boats (refer to Part 6 of BS 7671 and Chapter 10 of this book for clarification).
■ Ensure that exposed metalwork in 'supplier's works' that is not acting as a phase conductor is earthed, subject to minor exclusions set out in Regulation 8(2).

Part III – Electric lines below ground

Regulations 9, 10 and 11 cover underground cables and require such cables to have an earthed, continuous metallic screen enclosing the conductors, which may be used as the neutral. The intention is to protect a careless excavator from injury should the screen be penetrated with a tool and make contact with a phase conductor. The cables have to be buried deep enough to avoid damage from land use, and high voltage cables have to be further protected by pipes, ducts, cable tiles or a warning tape or device to alert anybody carrying out excavation work above the buried cable. Although this protection is not proof against a carelessly wielded pneumatic drill, it is otherwise fairly effective. Note that the requirement does not extend to low voltage cables, which account for many of the burn accidents suffered by excavators when they damage such cables.

Part IV – Electric lines placed above the ground

Regulations 12 to 16 are about overhead lines and mainly about safety clearances from the ground or from buildings and structures. The minimum heights above ground of the phase conductors of uninsulated overhead are set out in Schedule 2 and are shown in Table 8.1. The Regulations stipulate that high voltage lines must be protected against access and have safety signs attached to them. The Regulations apply to all owners of overhead supply lines.

Table 8.1 Minimum heights of phase conductors of uninsulated overhead		
Voltage of overhead	**Minimum height above road accessible to vehicular traffic (m)**	**Minimum height above ground other than road accessible to vehicular traffic (m)**
Not exceeding 33000 volts	5.8	5.2
Greater than 33 kV and less than or equal to 66 kV	6.0	6.0
Greater than 66 kV and less than or equal to 132 kV	6.7	6.7
Greater than 132 kV and less than or equal to 275 kV	7	7
Greater than 275 kV and less than or equal to 440 kV	7.3	7.3

Part V – Supplier's Works

Sufficiency of supplier's works

Regulation 17 is a general but onerous requirement that supplier's works be so designed, constructed, installed, protected, used and maintained as to prevent danger or interruption of the supply.

Maximum voltage

Regulation 18 imposes a maximum limit of 440,000 V on the supply voltage.

Enclosed spaces

Regulation 19 requires the supplier to avoid danger from the ingress of water or noxious or explosive liquids and gases into any enclosed space containing his works.

High voltage: additional provisions

Regulation 20 refers to the need to fence off high voltage substations and switching stations and similar premises to prevent unauthorised access. The height of the fence must be at least 2.4 m. Safety signs and notices must also be attached and precautions must be taken against fire.

Protective measures and precautions against excess voltage and supply failure

Regulation 21 requires excess current protection and evidently includes fault currents. Precautions against excess voltage are in Regulation 22 and against supply failure in Regulation 23.

Inspections

Regulation 24 requires the supplier to inspect his installations and works to ensure compliance with the Regulations.

Part VI – Supply to consumers' installations

Regulations 25 to 32 concern supplies to consumers' installations.

The requirements relating to the service line, protective device and wiring to the supply terminals are referred to in Regulation 25. Regulation 25(4) is almost a repetition of Regulation 19 and requires the entries of underground service cables to be sealed to prevent the influx of noxious gases and liquids. Water, however, is not mentioned nor is there any requirement to seal cable exits where, for example, a supply cable is looped into adjoining property. There are no specific requirements for the entry of an overhead service line. Regulation 25(5) as amended by the Statutory Instrument in 1990 requires the identification of low voltage conductors only at supply terminals, although Regulation 30(1) requires the supplier to 'declare' not just the supply voltage but also its frequency and the number of phases and their rotation.

Regulation 26 prohibits consumers from operating their own generators in parallel with the supply unless they have a written agreement with the supplier to do so. The connection of provide generators on to the public supply network is one area that has seen rapid growth in recent years, so much so that this is one aspect of the Regulations that needed to be updated to reflect current commercial practice and this is reflected in the new Electricity Safety, Quality and Continuity Regulations.

The standard of construction of the installation is required to match that referred to in Regulation 27, which stipulates that the supplier must satisfy himself before giving a supply to the installation that the installation is safe and will not cause undue interference. Regulation 27(2) says that any installation complying with the IEE Regulations (BS 7671) will be deemed to comply with this requirement. Suppliers will usually connect on receipt of a copy of the contractor's BS 7671 Completion and Inspection Certificate.

Regulations 28 and 29 detail the procedures for disconnection and provision of arbitration to settle any dispute between the supplier and consumer.

The declared phases, frequency and voltage and permitted frequency and voltage variation are in Regulation 30, but there is no requirement to provide a sine wave supply for a.c. or any limitation on ripple for d.c. supplies.

Regulation 31 requires the supplier to provide a written statement of the maximum prospective short circuit current at the supply terminals, the maximum earth loop impedance and the type and rating of the supplier's fuses or circuit breaker nearest to the supply terminals to anyone having reasonable cause for requiring this information. This Regulation could have been improved by requiring the supplier to notify the consumer(s) should he alter his network, so varying the parameters in which the consumers' equipment is endangered.

Regulation 32 emphasises the duty that electricity suppliers have to ensure continuity of supply except in exceptional circumstances such as accidents occurring on the network.

Part VII – Miscellaneous

Regulation 33 empowers the Secretary of State's inspectors to carry out inspections and tests of a supplier's works and instruments. Accidents and dangerous occurrences are notifiable under Regulation 34 and supply failures under Regulation 35.

Underground cable plans are dealt with under Regulation 36. This is an important provision in the context of reducing the number of occurrences of underground cables being struck. The supply companies must maintain maps indicating the position and depth of cables and other elements of the distribution system and must make the maps available to anyone who can show reasonable cause. This includes local planning authorities and construction companies and utilities such as gas and telecommunication system suppliers. Some of the supply companies now hold their maps in electronic form and make them available via the internet.

Requests for exemptions are dealt with in Regulation 37 and enforcement procedures in Regulation 38. Regulation 38(1)(a), (b) and (c) curiously appears to cover suppliers' works and consumers' outdoor but not indoor installations unless the latter are in breach of an exemption.

Regulation 39 as amended increases the maximum penalty for an infringement from level 3, on the standard scale, to level 5.

Schedules

The Schedules deal with the design of the safety sign, heights above ground of overhead lines, conditions for parallel operation, and notification of accidents, dangerous occurrences and supply failures.

THE ELECTRICITY SAFETY, QUALITY AND CONTINUITY REGULATIONS 2002

As previously noted, the draft Electricity Safety, Quality and Continuity Regulations were distributed for comment in February 2001. They are likely to be promulgated in the middle of 2002, and the following describes the main requirements and the principal changes from the 1988 Regulations as deduced from the consultation documents.

The technical content of the new Regulations concerning matters such as earthing and protection tends to match that of the 1988 Regulations, although it is written in a much less prescriptive and detailed fashion. The substantial changes cover the allocation of duties to those entities now operating in the electricity supply market. From the perspective of safety, there are also important new requirements concerning the need for duty holders to assess and control the risk of injury from plant and overhead lines arising from interference, vandalism and unauthorised access, and from inadvertent damage to low voltage underground cables during excavation work.

The DTI Engineering Inspectorate is drafting a guidance document to explain the regulations and it is anticipated that this will become available when the regulations are published.

Scope of the new Regulations

In similar vein to the 1988 Regulations, the new Regulations specify safety standards aimed at protecting the general public and consumers from danger. They also specify power quality (in terms of voltage and frequency limits) and supply continuity standards. The Regulations are made by the Secretary of State for Trade and Industry under the Electricity Act 1989, taking account of amendments made by the Utilities Act 2000.

In recognition of the significant changes that have taken place in the supply industry, the duty holders are no longer limited to electricity boards. Licensed and non-licensed generators, distributors, suppliers and meter operators, as well as their respective agents and contractors, are all included within the scope of the proposed legislation. Private networks are covered as well as those used for supplies to the general public.

The main requirements of the Regulations and changes from the 1988 Regulations

Part I

Part I includes a list of definitions. Many are the same as the 1988 Regulations, although the number of definitions has been reduced. The most important of the new definitions concern the new type of duty holder and are as follows:

(1) *Distributor* A distributor is an organisation, most obviously an electricity supply company but it may also include other organisations and individuals, that owns or operates a network, where a 'network' is a transmission or distribution system that supplies electricity to connected consumers. Note that this excludes networks belonging to railway companies for supplying electric trains and networks situated entirely offshore.

(2) *Meter Operator* This term describes 'a person', which will usually mean a company, who owns, installs, maintains or removes metering

equipment; this refers to meters used for measuring the amount of energy flowing (typically for billing purposes) rather than metering that is used for electrical protection or other purposes. Note that an organisation such as a 'distributor' may also fall within the scope of being a 'meter operator', in which case the organisation will be captured by the requirements for both types of duty holder.

(3) *Generator* This term refers to generation companies who generate electricity and supply it to networks for transmission and distribution to connected consumers. Since it only relates to generators supplying networks, it can be concluded that the term does not refer to generators that form part of, for example, domestic or small-scale commercial/industrial installations.

(4) *Supplier* A supplier is a person or organisation who contracts to supply electricity and who may be involved with the reading of electricity meters. For example, a gas company that purchases and sells electricity may contract to supply electricity to a domestic property, despite the fact that the company does not generate or distribute electricity. The gas company would be a 'supplier'.

Regulation 2 describes when certain provisions of the Regulations have to be applied. These will be covered later in the chapter when the appropriate regulations are described.

Regulation 3 is a catch-all regulation specifying that duty holders must ensure that their equipment is 'sufficient' for its purposes and circumstances, and that it is constructed, installed, protected, used and maintained so as to prevent danger or interruption of supply, so far as is reasonably practicable.

Regulation 3(2) stipulates that generators and distributors have to maintain a risk/action register relating to unlawful interference, vandalism or unauthorised access for each substation and overhead line support. Given the number of injuries that members of the public suffer as a result of such behaviour, the need for this new requirement is understandable and is to be welcomed. This is a new and important requirement that will require risk assessments relating to the potential for interference, vandalism or unauthorised access to be carried out and for appropriate action to be taken to reduce the risk to an acceptable level. The assessments will need to take account of factors such as the dangers posed by the particular equipment (such as direct contact with overhead power lines); the geographic and social environment in which the equipment is located; and the history of incidents of equipment and plant being interfered with, vandalised or accessed by the likes of children and people intent on stealing property. Actions that could be taken include, for example, replacing uninsulated overhead lines in high risk areas with underground cables or insulating the overhead lines; and improving the effectiveness of substation security fencing in high risk areas. However, the cost of the remedial measures will need to be weighed against the benefits, using the form of reasonable practicability calculations explained in Chapter 4. Compliance in the case of overhead lines must be secured within five years of the Regulations being published, and within two years in the case of substations.

Regulation 5 stipulates that networks must be regularly inspected and that, in the case of substations and overhead lines, records of the inspections and any recommendations for action arising from them have to be maintained for at least ten years.

Part II

Part II addresses protection and earthing arrangements. The protection requirements are non-prescriptive and relate to precautions, such as the provision of fuses and circuit breakers, against excess current and earth leakage currents causing damage.

The regulations covering continuity of neutral conductors, connection with earth, protective multiple earthing, and earthing of metalwork are significantly less detailed and prescriptive than the 1988 regulations; for example, the maximum resistance to earth of the supply neutral conductor in PME supplies, and the copper equivalent cross sectional area of bonding conductors, are no longer specified. This goal-setting approach is consistent with other legislation covering safety matters. Technical requirements relating to low voltage installations are covered in BS 7671 (see Chapter 10), to which these regulations specifically refer (in regulation 25) when stipulating that connections to consumers' installations must not be made unless they comply with BS 7671.

Regulation 7(2) requires that no protective device (such as a fuse) must be installed in the neutral or earthing conductor in low voltage installations. One of the effects of this is to outlaw pre-1937 fusible cut-outs in which there is a fuse in the neutral conductor; these cut-outs must be removed from the system within 10 years of the Regulations coming into force.

Part III

Part III contains requirements relating to substations, mainly in the context of preventing access to them by having suitable enclosures or fencing. For example, open air substations that include live uninsulated equipment, such as open busbars and terminations, must be enclosed with a fence that is at least 2.4 m high. Other requirements cover the displaying of triangular 'Danger of Death' safety signs; a notice which identifies the substation (by location or name), the name of the generator or distributor controlling the substation, and a valid contact telephone number; and any other signs that may be needed to reduce the risk of injury. The requirements for the displaying of signs must be complied with within two years of the Regulations being enacted.

Part IV

Part IV covers requirements for underground cables and equipment. They mirror the requirements of the 1988 Regulations, with one notable exception. Regulation 14(2) has the welcome requirement that all buried cables must be marked or otherwise protected. This would extend the requirements currently only relating to high voltage cables to low voltage cables, meaning that cables must be placed in pipes or ducts or be overlayed with cable tiles or warning tape or some other means of indicating the presence of the cables. The requirement is not retrospective for low voltage cables but, nonetheless, hopefully this will eventually result in a reduction in the unacceptably high number of cables that are struck during careless excavation work.

Regulation 15 covers the very important duty placed on generators and distributors to produce and maintain maps of their underground service cables, showing their position and depth. These must be made available in either paper or electronic form to certain people on request, including planning authorities and those who can show reasonable cause. This would include, for example, civil contractors who are carrying out excavation work and want to check for the presence of buried cables before starting to dig. Checking such maps is a crucial element of safe digging practice to prevent damage to buried cables.

Part V

Part V provides requirements relating to overhead lines. They again mirror the requirements of the 1988 Regulations, particularly with regard to the minimum heights of overhead lines above ground. A very significant proposal in the first consultation round had been that new overhead lines and existing lines that are refurbished or rebuilt would have to be at a height of at least 7.3 m over roads and other surfaces designed to be used by vehicles. This compares with the current 5.8 m for 11 kV overhead lines and was aimed at reducing the number of occurrences of overhead lines being struck by vehicles. However, this change did not make it through the consultation process and the heights prescribed in the 1988 Regulations have been retained.

Regulation 19 requires 'Danger of Death' safety signs to be fitted to high and low voltage overhead line supports within ten years of enactment of the

Regulations, but only if the lines are uninsulated. It also stipulates that devices should be fitted to high voltage overhead line supports to prevent unauthorised access to the live conductors – typically, this would be barbed wire strands surrounding the legs of the supports to deter people from climbing them. Similarly, Regulation 20 stipulates that, within the same ten year timeframe, stay wires on uninsulated-line supports must be fitted with an insulator no lower than 3 m above ground level or above the normal height of the line.

Part VI

Part VI covers embedded generation systems, operating as either switched sources of energy or in parallel with a distributor's network. It has been included in recognition of the increasing use of embedded generators in consumers' premises. Low voltage installations would have to comply with BS 7671, which contains requirements in section 551 relating to generation systems (see Chapter 10). Generators paralleled onto the network which have power outputs less than 5 kW would be excluded from the scope of the regulations, subject to certain provisions.

Part VII

Part VII addresses the supplies to installations and covers matters such as precautions against supply failure; the suitability and maintenance of distributor or meter operator equipment on consumers' premises; the circumstances under which supplies can be connected and disconnected; and the information that must be provided to consumers. A new requirement in Regulation 24 would mean distributors having to offer earth facilities with new connections at low voltage. Moreover, under Regulation 25, the distributor should not provide a connection if there are reasonable grounds to believe that the consumer's installation, or other distributor's network, does not comply with BS 7671 or other relevant standards.

There has been a new addition to the information to be provided by distributors free on request to low voltage consumers. The 1988 Regulations stipulate that they must provide a written statement of maximum prospective short circuit current; maximum earth loop impedance; and the type and rating of supplier's fusible cut-out or switching device. The new Regulations add to this the type of earthing system (TT, TN etc).

Part VIII

Part VIII covers a range of miscellaneous provisions similar to those in the 1988 Regulations. These include requirements concerning inspections by

DTI engineering inspectors; the notification of specified events (such as death caused by systems covered by the Regulations) and certain interruptions of supply; exemptions from the Regulations; breaches and offences; and the revocation of the previous Electricity Supply Regulations and the Electricity (Overhead Lines) Regulations 1970.

Schedules

The Schedules deal with the design of the 'Danger of Death' safety sign, heights above ground of overhead lines, notification of events and interruptions of supply, and revocations of other legislation.

Chapter 9
Other Legislation with an Electrical Safety Content

THE PETROLEUM (CONSOLIDATION) ACT 1928

The Petroleum (Consolidation) Act 1928 originally applied only to the storage and dispensing of petroleum spirit, but has been extended by a number of Orders to include a long list of other flammable liquids. The Act was produced by the Home Office but is now the responsibility of the HSE. It is generally enforced by local authorities who license the installations for the keeping of petrol and other substances. Harbour authorities cover filling stations in harbour areas, and the HSE have responsibilities where a site is subject to the Notification of Installations Handling Hazardous Substances Regulations 1982.

Comprehensive guidance is now available in three documents. The technical criteria to be adopted in petrol filling stations, including for the electrical parts of the installation, are published in an industrial guidance booklet published by the Association for Petroleum and Explosives Administration (APEA) and the Institute of Petroleum (IP) titled *Guidance for the design, construction, modification and maintenance of petrol filling stations*. The guidance in this booklet supersedes the technical guidance previously published in HSE guidance booklets HS(G)41 *Petrol filling stations: Construction and operation* and HS(G)51 The storage of flammable liquids in containers. The APEA/IP guidance is not retrospective, so it applies to new and refurbished petrol filling stations.

This guidance book, which runs to 241 pages is very thorough and informative, providing advice on safety during the life cycle of a petrol station, from inception through to decommissioning. Part 14 covers the electrical installation and is essential reading for anybody involved with the specification, design, commissioning and maintenance of the electrical parts of filling stations.

The content of HS(G)41 relating to the operation of a petrol filling station remains current, although it is intended that updated guidance will be published in due course. However, the HSE has published Guidance Note

HS(G)146 Dispensing petrol: Assessing and controlling the risk of fire and explosion at sites where petrol is stored and dispensed as a fuel. The purpose of this guidance is, perhaps, self-explanatory from its title.

EXPLOSIVES ACT 1875

Commercial explosives are subject to the Explosives Act 1875 which is enforced by the HSE's Hazardous Installations Directorate for factories and magazines, by local authorities for explosive stores (up to 2 tons) and for registered premises for retail sale. All such premises are licensed. Similar controls apply to military explosives which are enforced by the Ministry of Defence. The Act requires precautions to be taken to prevent accidental ignition, which is interpreted by the enforcement authorities to mean the use of explosion-protected installations similar to those in other industries with flammable hazards, together with efficient lightning protection and stringent antistatic measures.

CINEMAS

The Cinematograph (Safety) Regulations 1955 and subsequent amendments lay down the requirements for the electrical installation in cinemas. Where both the general and safety lighting are electric, there have to be two sources of supply to ensure maintenance of the emergency lighting in the event of mains failure. These Regulations are enforced by the local authorities.

SCOTTISH BUILDING REGULATIONS

The Building Standards (Scotland) Regulations 1990 came into force on 1 April 1991. They have been amended since then but the amendments have not affected the electrical installation requirements in Regulation 26, which call for the installation to be adequately protected so that it is not a source of fire or personal injury. Regulation 9 requires the installation to meet Technical Standards, i.e. British Standards and/or CENELEC (European Committee for Electrotechnical Standardisation) harmonised documents. Part N of the Regulations stipulates that compliance with the 16th Edition of the IEE Wiring Regulations, BS 7671, will secure compliance with requirements to ensure that electrical installations are safe.

The Regulations are enforced by Building Control Departments of local authorities in Scotland, who generally require a BS 7671 completion certificate before they will issue building warrants for new or refurbished buildings that fall within the scope of the Regulations. Currently, this is the only legal requirement relating to BS 7671, although there is continuing consultation about incorporating a similar requirement in the English and Welsh version of these Regulations.

Chapter 10

BS 7671 Requirements for Electrical Installations

INTRODUCTION

The first edition of the Institution of Electrical Engineers (IEE) Wiring Regulations for electrical installations was published in 1882, under the title 'Rules and Regulations for the Prevention of Fire Risks Arising from Electric Lighting'. Since then these have acted as the benchmark design standard for ensuring that low voltage installations do not jeopardise the safety of persons and livestock. The 16th edition of the Regulations was published in May 1991, superseding the 15th edition which had been published in 1981. In 1992 it was adopted as a British Standard, BS 7671 : 1992 Requirements for Electrical Installations, although the document remains generally known as 'the IEE Wiring Regulations'. The standard has been amended on four occasions since 1992 to take account of additions and amendments to the IEC publications and CENELEC standards on which it is now largely based; these amendments will be outlined later in this chapter. When the fourth amendment was issued in July 2001, all the previous amendments were incorporated in a reprint and the standard was updated to BS 7671 : 2001.

The standard is the British version of IEC Publication 364 *Electrical Installations of Buildings* and incorporates the provisions of CENELEC harmonised standards, as considered and agreed by the joint IEE/BSI Technical Committee JPEL/64.

SCOPE

The standard establishes the accepted safety parameters for the designers, erectors and testers of low voltage installations not exceeding 1000 V a.c. and 1500 V d.c. There is a list of instances, such as installations in flammable atmospheres, where the standard needs to be supplemented by the requirements of more specific British Standards. There is also a list of exclusions:

- 'supplier's works', as defined in the Electricity Supply Regulations 1988;
- railway systems;
- motor vehicles (although there are requirements relating to caravans and caravan parks);
- equipment on offshore platforms and ships (the IEE publishes a set of regulations that are specific to maritime applications);
- equipment on aircraft;
- equipment in mines and quarries where covered by other Statutory Regulations;
- radio interference suppression equipment (unless it affects the safety of electrical installations), lightning protection equipment (refer to BS 6651), and those aspects of lift installations covered by BS 5655.

STATUS

BS 7671 has the status of a British Standard code of practice. The term 'IEE Wiring Regulations' should not be misinterpreted as meaning that the standard is a statutory instrument. However, the standard is recognised as the benchmark for the design of low voltage electrical installations in the UK. In Scotland, the Building Standards (Scotland) Regulations 1990 require that all low voltage electrical installations in buildings comply with the standard. At present in the UK this is the only legal reference to the standard, although the DETR (now the Department for Transport, Local Government and the Regions – DTLR) has been considering for some time whether or not to include a similar requirement in the next revision to the England and Wales version of the Building Regulations.

The Memorandum of Guidance to the Electricity at Work Regulations refers to the Standard and notes that installations that are designed in conformity with it are likely to achieve compliance with the Electricity at Work Regulations 1989. The HSE's electrical inspectors will generally use BS 7671 as a main guide when checking the safety of electrical distribution systems.

The standard is also commonly used in civil litigation concerning the safety of consumer's electrical installations.

ENFORCEMENT

The only enforcing authority for the standard is the Building Control Officers in Scotland, who seek to ensure compliance with the Building Standards (Scotland) Regulations, and who will therefore not issue completion certificates for new buildings unless the electrical installation conforms with BS 7671. This situation would change should use of the standard become a statutory requirement in the Building Regulations in England and Wales, a matter that has been under consideration for a considerable time.

Many electrical contracting companies who install electrical distribution systems in conformity with BS 7671 are registered with the National Inspection Council for Electrical Installation Contracting (NICEIC), a United Kingdom Accreditation Service (UKAS) accredited organisation concerned with ensuring the safety of electrical installations. Some are also members of the industry trade associations: the Electrical Contractors' Association (in England and Wales) and SELECT (in Scotland). All three organisations periodically inspect the work of relevant companies against the standards laid out in BS 7671. The ultimate penalty for non-compliance is deregistration.

GUIDANCE

The standard can be quite difficult to follow and to interpret, so it is helpful that the IEE has produced a series of explanatory guidance notes to support it. The following guidance notes are published as separate documents:

- Guidance Note 1: Selection and Erection.
- Guidance Note 2: Isolation and Switching
- Guidance Note 3: Inspection and Testing
- Guidance Note 4: Protection against Fire
- Guidance Note 5: Protection against Electric Shock
- Guidance Note 6: Protection against Overcurrent
- Guidance Note 7: Special Locations

AMENDMENTS

There have been four amendments to the standard since it was first published.

First amendment

The first amendment was issued in December 1994 to become effective on 1 January 1995. It had the main purpose of reflecting the harmonised low voltage ranges that had been agreed by CENELEC. The aim of this was to harmonise low voltage mains electricity supplies in Europe at 230 V. Thus, whereas prior to 1 January 1995 the permitted voltage band in the UK was 240 V +6%/−6%, after 1 January 1995 it was 230 V +10%/−6%. From January 2003, it is proposed that the bands will be 230 V +10%/−10%.

Second amendment

The second amendment was issued in December 1997 to become effective on 1 January 1998. The amendment:

- Required SELV circuit conductors to be insulated and sheathed (section 411-02-06(ii)).
- Altered the SELV test voltage from 500 V d.c. to 500 V a.c. (section 411-02-09(ii)).
- Permitted 30 A ring mains to be wired with 1.5 mm² 2 core MIMS cables (section 433-02-04).
- Required the neutral, in single-phase circuits, to have a cross-sectional area of not less than the phase conductor and, in polyphase circuits, the neutral to be of adequate size for the anticipated current (section 524-02-01).
- Required RCDs, operated by unskilled persons, to be unadjustable without the use of a key or tool (section 531-02-10).
- Inserted a new section entitled 'Generating Sets'. It embraces stand alone sets and stand-by sets (both permanent and portable) and deals with the protection required and agreement needed with the supply company where stand-by sets are used (section 551).
- Inserted new requirements for the provision of a Minor Electrical Installation Work Certificate to cover work which does not include a new circuit (sections 743-01-02 to 04).

Third Amendment

The third amendment became effective in April 2000. It was a complete revision of section 601, which had been entitled 'Locations Containing a Bath Tub or Shower Basin' and which became, in the amendment, 'Locations Containing a Bath or Shower'.

The requirements in the amended section 601 relate to locations containing baths, showers and cabinets containing a shower and/or bath. These can be domestic, commercial or industrial locations, although a new exclusion means that emergency facilities in industrial areas (such as decontamination showers) and laboratories are excluded.

There had been considerable debate during the preparation of this amendment about a proposal to cancel the long-standing prohibition on the presence of standard (IP2X) mains socket outlets in bathrooms. The proposal was that continental practice should be adopted to allow socket outlets to be installed in bathrooms, subject to restrictions on their location and a requirement that they must be protected by a residual current device. In the final analysis this proposal was rejected, principally on the grounds that since RCDs can fail to danger, with unrevealed faults, they should not be the main protection against electric shock in wet environments.

The major change introduced by the amendment was the concept of four zones, ranging from Zone 0 to Zone 3, which delineate the bath or shower room and in which the minimum ingress protection properties of installed electrical equipment are specified. The zones are illustrated in Fig. 10.1 for the straightforward situation of a bath tub in a bathroom. Reference should be made to the standard for the distribution of zones in more complicated situations. The most important requirements relating to these zones are detailed in Table 10.1.

There are also provisions in section 601-04 relating to supplementary equipotential bonding. This requires the terminals of protective conductors associated with Class I and Class II equipment in Zones 1, 2 and 3 to be connected together and for them to be connected to extraneous conductive parts on those zones. The latter parts are listed as including metallic items such as service pipes (gas, water and so on), waste pipes, central heating pipes, air conditioning duct work, accessible structural parts of buildings, and baths and shower basins.

This requirement for supplementary bonding is frequently misunderstood. Although the exposed metalwork should already be earthed by the main equipotential bonding, additional safety precautions are needed to make sure that all touchable metalwork is at the same potential because of the increased risk of electric shock in the wet environment of the bathroom. Connecting together the metalwork is called 'bonding', and this is done using a 2.5 mm² sheathed conductor with green/yellow insulation (4 mm² if mechanical protection is not provided). It is called 'supplementary' bonding because it supplements the main equipotential bonding. A typical supplementary bonding configuration is illustrated in Fig. 10.2.

The amendment recognises that it is common for shower cubicles to be installed, for example as en-suite facilities in bedrooms. Under those circumstances, section 601-08-02 stipulates that 230 V standard socket outlets must be installed outside zones 0, 1, 2 and 3 and must be protected by an RCD with a rated operating current not exceeding 30 mA.

Fourth amendment

The fourth amendment was issued on 1 June 2001, and came into effect on 1 January 2002. As well as incorporating the first three amendments and

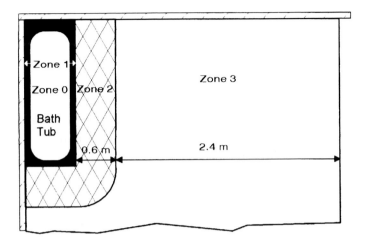

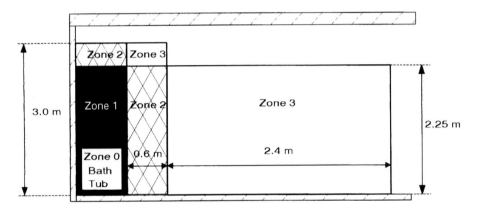

Fig. 10.1 Example of zone dimensions.

updating the standard from its 1992 edition to a 2001 edition, it introduced a number of additional requirements, the main ones being summarised here.

Part 1 was expanded to give a deeper explanation of the standard's scope, objectives and fundamental principles.

Section 443 was introduced to provide requirements for protection against transient overvoltage of atmospheric origin, mostly lightning strikes on the distribution network, or due to switching on the network. The main objective is to ensure that installations that contain equipment with inadequate impulse withstand voltage, or which are supplied via uninsulated overhead lines and which are subject to frequent thunderstorm activity, have surge protection against transient overvoltages. Note that this requirement is

Table 10.1 Zone requirements

ZONE 0 REQUIREMENTS

■ Electrical equipment in the zone must have the degree of protection of at least IPX7. Only equipment which can reasonably only be located in the zone is permitted and it must be suitable for the conditions of use in the zone. Moreover, no switchgear or accessories are permitted unless they are incorporated in fixed current-using equipment suitable for use in the zone.
■ The only permitted protection against electric shock is SELV at a nominal voltage not exceeding 12 V rms a.c. or 30 V ripple-free d.c. so long as the safety source is outside zones 0, 1 and 2.
■ Only wiring systems feeding fixed electrical equipment in the zone are permitted.

ZONE 1 REQUIREMENTS

■ Electrical equipment in the zone must have the degree of protection of at least IPX4. However, where water jets are likely to be used for cleaning purposes in communal baths or showers, a degree of protection of at least IPX5 must be used.
■ Only wiring supplying fixed electrical equipment located in zones 0 and 1 must be installed.
■ Only switches of SELV systems at a nominal voltage not exceeding 12 V rms a.c. or 30 V ripple-free d.c. are permitted so long as the safety source is outside zones 0, 1 and 2.
■ The following equipment may be installed so long as it is suitable for the zone:
  – Water heaters
  – Shower pumps
  – Other fixed current-using equipment which can reasonably only be located in zone 1, but it must be suitable for the conditions of use, and the supply circuit must be additionally protected by an RCD with a rated residual operating current of 30 mA.
  – SELV equipment.
  – The insulated cord of cord-operated switches complying with BS 3676.

ZONE 2 REQUIREMENTS

■ Electrical equipment in the zone must have the degree of protection of at least IPX4. However, where water jets are likely to be used for cleaning purposes in communal baths or showers, a degree of protection of at least IPX5 must be used.
■ Only wiring supplying fixed electrical equipment located in zones 0, 1 and 2 must be installed.
■ Switchgear, accessories incorporating switches or socket outlets must not be installed with the exception of:
  – switches and socket outlets that are part of SELV circuits in which the safety source is not inside zones 0, 1 and 2; and
  – shaver supply units complying with BS EN 60742 Chapter 2, section 1.
■ The following equipment may be installed if it is suitable for use in zone 2:
  – Water heaters
  – Shower pumps
  – Luminaires, fans, heating appliances and units for whirlpool baths complying with appropriate standards.
  – Other fixed current-using equipment which can reasonably only be located in the zone and which is suitable for use in it.
  – SELV equipment.
  – The insulated cord of cord-operated switches complying with BS 3676.

ZONE 3 REQUIREMENTS

■ Electrical equipment that is likely to be subjected to water jets must be protected to IPX5, otherwise there are no particular ingress protection requirements.
■ SELV socket outlets and shaver units to BS EN 60742 Chapter 2 section 1 are permitted. Other accessories are permitted, but there must be no other socket outlets.
■ Fixed equipment is permitted, and this may be 230 V equipment.
■ Any other equipment that is not fixed must be protected by an RCD with a rated operating current not exceeding 30 mA.
■ The insulated cord of cord-operated switches complying with BS 3676.

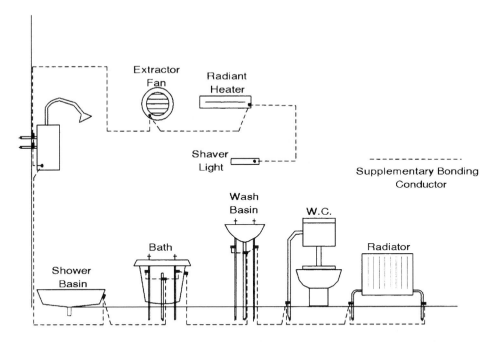

Fig. 10.2 Supplementary bonding of touchable metalwork in a bathroom.

unlikely to apply to most places in the UK, where thunderstorm activity can generally be taken to be below the threshold set in the standard.

There are new requirements in section 473 relating to overcurrent protection of conductors connected in parallel.

Section 482 was introduced to recommend precautions where there is a particular risk of, or danger from, fire. The section's title refers to 'risks of danger', which is a bit tautological since 'danger' is often defined in the electrical safety world as the 'risk of injury'. Nonetheless, its main thrust is to minimise the risk of fire in installations where there may be combustible materials being processed or stored (such as sawmills and paper mills), or where the building is constructed of combustible materials. The main precautions are summarised as:

■ only installing electrical equipment in such areas when necessary;
■ restricting the surface temperature of enclosures and luminaires and the running temperatures of motors;
■ ensuring enclosures have an adequate IP rating to exclude dust (IP5X);
■ having wiring systems with suitable flame propagation characteristics and adequate protection against mechanical damage;
■ providing for the detection and disconnection of earth faults;

■ restricting the use of combined neutral and protective (PEN) conductors;
■ specifying the location and construction of spotlights, projectors and lamps to minimise the chance of them acting as ignition sources;
■ using cables, cords, conduits and trunking that have suitable fire resistance and fire propagation properties.

Section 607, dealing with earthing requirements on systems incorporating equipment with high protective conductor currents, was substantially rewritten, although the detailed technical requirements did not change to any great extent. Note the use of the term 'protective conductor currents' which was used for the first time in this edition to replace the previously-used term 'earth leakage current'. The guidance is aimed at installations that contain large quantities of information technology equipment (call centres, financial trading centres and the like), equipment incorporating radio frequency suppression filters, and heating elements. These forms of equipment can lead to substantial currents flowing in protective conductors under normal conditions – this can lead to hazardous situations, particularly if the protective conductor were to open circuit and impress substantial voltages between exposed metalwork and earth. The section provides detailed requirements concerned with maintaining the integrity of the protective conductors and ensuring that they have adequate cross sectional area for the prospective current.

PART 1 SCOPE, OBJECT AND FUNDAMENTAL PRINCIPLES

Chapter 11: Scope

The Regulations in this chapter lay out the broad scope of the standard, as already explained above. Regulation 110-04-01 advises that adherence to the standard is also intended to ensure compliance with statutory requirements. There is a list of the relevant statutory Regulations in Appendix 2 of the standard.

Chapter 12: Object and effects

This chapter now simply states that the standard aims to provide systems that are safe and which function properly. Regulation 120-02-01 states that invention, and new materials which may infringe the standard, are not prohibited provided they attain a comparable degree of safety and their use is noted on the completion certificate.

Chapter 13: Fundamental principles

Section 130-01-01 clarifies that the purpose of the standard is to protect persons, property and livestock against the injurious effects of electric shock, burns, mechanical movement of electrically actuated equipment, and explosion. So only trained, competent and conscientious operatives should be employed and the materials should be suitable and of good quality to the appropriate British or other standard.

In many ways, the use of the term 'so far as is reasonably practicable' in the large majority of the Regulations in section 130 demonstrates that this is a rather legalistic section, setting out the standard's foundations upon which the more prescriptive Regulations in later chapters are built. Contractors, however, should not overlook section 130-07-01 when extending or altering an installation, as upgrading work on an old installation will often be needed to meet the earthing requirements. In older domestic accommodation, for example, where the supply system is often TT, a residual current device (RCD) will be required to feed socket outlet circuits (see section 471-08-06). Also, any old voltage-operated earth-leakage circuit breaker will need to be replaced by a residual current circuit breaker.

Sections 131 and 132 give advice on the general parameters that must be taken into account when designing an installation and selecting equipment. This does not cover simply the electrical characteristics of the installation (voltage, frequency, nature of demand and so on) but also the very important factor of the environmental conditions in which the installation is to be used. For example, the highly corrosive atmosphere of an electroplating factory will demand very different design solutions from those in the installation in a cinema where large numbers of members of the public congregate.

Section 133 addresses, among other things, the need for installations to be verified by using inspection and testing to ensure that the provisions of BS 7671 have been met.

PART 2 DEFINITIONS

Part 2 provides an extensive list of definitions of terms used throughout the standard. One definition that needs to be clarified is that of the earthing arrangements. There are four descriptors in general use which are used both in BS 7671 and in this book. They are TN-S, TN-C-S, TT and IT systems. In these abbreviations, the letters have the following meanings:

First letter:
T: Direct connection of the source with earth at one or more locations

I: All source live parts isolated from earth, or connected with earth through high impedance.

Second letter:

N: Exposed conductive parts connected directly by protective conductor with source earth.

T: Exposed conductive parts connected by protective conductor with source earth via an installation earth electrode.

Third letter:

C: Single conductor provides for both neutral and protective conductor functions.

S: Separate conductors for neutral and protective conductor functions.

C-S: Neutral conductor and protective conductor combined in the supply and separated in the installation.

The TN-S, TN-C-S and TT systems, which are by far the most common systems in the UK, are depicted in Fig. 10.3 for a single-phase supply. The fuse in the diagrams is the supply company's fuse on the supply to the typical domestic installation, usually rated at either 60 A or 100 A.

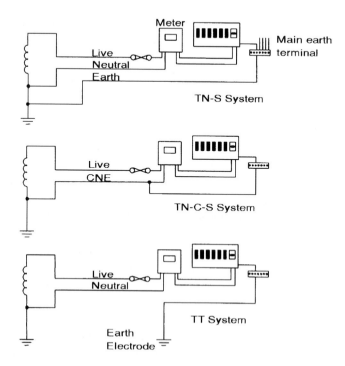

Fig. 10.3 Earthing systems on single-phase low voltage supply.

PART 3 ASSESSMENT OF GENERAL CHARACTERISTICS

Part 3 lists the initial factors that the designer should consider when planning an installation, to ensure its viability and safety. It is usually essential to contact the electricity supply company early and advise the total load and possible future increases so that an adequate supply will be available when required. Spaces for substations, standby facilities and switchrooms, and any cable duct requirements should be discussed with whoever is responsible for the site and buildings as soon as possible so that provision can be made in the plans.

Regulation 314-01-04 requires a separate way for each final circuit in a distribution board, which includes a consumer unit, and Regulation 314-01-02 calls for separate circuits where required by the Regulation or to prevent danger. In this respect the user should be consulted to identify the hazards. Where a lighting failure could cause danger, for example, there should be more than one lighting circuit, or emergency lighting may be appropriate. Although not mentioned in the Regulation, it is advisable to provide spare circuits for future use.

The environmental conditions and other external influences, listed in Appendix 5 of the standard, should be considered to see if any apply and what should be done to counter them. For example, in rural areas the electricity supply might be an overhead line in a lightning-prone zone where it may be prudent to install voltage surge suppressors (see new section 443) and perhaps fireproof the supply intake. In nurseries and schools, the installation should be child-proof and electric heaters located out of reach. Other factors affecting the selection of equipment are detailed in Part 5.

Connected equipment with harmful characteristics is referred to in Chapter 33 – Compatibility. The characteristics include transient over-voltages which may emanate from the supply or from switching inductive loads, harmonic currents, d.c. feedback, mutual inductance, high frequency oscillations, large starting currents and large fluctuating loads. Appropriate remedial action is required to prevent adverse effects, particularly to sensitive electronic apparatus such as computers and programmable electronic systems. Where there are no practicable remedies to counter large starting currents and the fluctuating loads of arc furnaces, for example, the supply company has to be advised so that it can make suitable alterations to its network and to avoid the risk of being denied a supply or being disconnected under the provisions of section 27(1) of the 1988 Electricity Supply Regulations.

The maintainability requirement of Chapter 34 entails the selection of durable equipment which can be readily opened for servicing, and its location in an accessible position. This is a requirement frequently overlooked by architects who have to shoehorn electrical equipment into modern space-limited buildings.

PART 4 PROTECTION FOR SAFETY

Part 4 lists the methods of providing protection against direct and indirect electric shock, thermal effects (fire, burns and overheating), excess of current, overvoltage and undervoltage in Chapters 41, 42, 43, 44 and 45 respectively. Chapter 46 addresses the requirements for isolation and switching. Chapter 47 and Part 6 explain how these measures should be applied, qualifying and amplifying the Regulations in the preceding sections.

In the following text, a direct shock is from contact with a live part which is intentionally live. An indirect shock is from contact with an exposed conductive part or an extraneous conductive part made live from a fault.

Section 411 Protection against both direct and indirect shock

Safety extra-low voltage (SELV)

The intention of SELV is to minimise the shock hazard by voltage limitation. Subsection 411-02 sets out the parameters, which include a safe source of supply such as a safety transformer to BS 3535 and other precautions to avoid the SELV circuit becoming live at a higher voltage. If the SELV does not exceed 25 V a.c. or 60 V ripple-free d.c., i.e. not more than 10% ripple, the direct contact shock risk is regarded as negligible and exposed live parts are allowed except in locations of enhanced shock risk, such as most of those in Part 6 where the conductors have to be protected against direct contact by a barrier, enclosure or insulation. The SELV circuit is not earthed and the cables are not metal-sheathed. As conductive parts of the installation are not deliberately or fortuitously earthed, it will usually be more practicable to use insulated rather than metallic conduit and ducts.

Other extra-low voltage systems

Where it is not possible to meet all the requirements in subsection 411-02 for SELV, but the voltage does not exceed extra-low voltage, the protection is called protective extra low voltage (PELV). The difference between a SELV and PELV system is that the latter is connected to earth but in other respects meets the SELV requirements. Other extra-low voltage systems are called functional extra-low voltage (FELV). The relevant precautions are detailed in section 471-14. The socket outlets in SELV, PELV and FELV systems have to be different from those used for higher voltages so that the plugs used to connect the apparatus cannot be inserted into a higher voltage socket outlet.

Reduced voltage system

Another method is known as the automatic disconnection and reduced low voltage system and is described in subsection 471-15. The maximum voltages are 110 V a.c. three-phase with earthed neutral or 110 V a.c. single-phase with mid-point earthing so that the phase-to-earth voltages are, respectively, 63.5 V and 55 V. Precautions are specified to ensure that these voltages are not exceeded. Protection against direct shock is by insulation, barriers or enclosures and against indirect shock by automatic disconnection of the supply in not more than 5 s in the event of an earth fault. The protective devices may be fuses or circuit breakers with overload trips or RCDs. To ensure operation within 5 s, the maximum earth fault loop impedances allowable for miniature circuit breakers (MCBs) and BS 88 fuses are given in Table 471A. Where the earth fault loop impedances exceed the tabulated figures, RCDs should be used to achieve the disconnection times. Again, the socket-outlets have to be different from those used for higher voltages. The main UK usage is for supplies to portable apparatus, particularly on construction sites and similarly harsh environments, as described in Chapter 11 of this book.

Protection by limitation of energy

Protection by limitation of energy is in subsections 411-04 and 471-03. The intention is not to prevent the shock sensation but to limit the shock current and/or its duration so as to avoid injury to persons and animals. Common examples of equipment complying with this requirement are electric fence energisers, electrostatic paint and powder sprayers, and tungsten inert gas (TIG) welding electrodes.

Avoiding direct contact

Measures to prevent shock by direct contact are in section 412 and subsections 471-04 to 471-07 and include the insulation of live parts, the use of barriers, enclosures and obstacles, and placing out of reach. The intention is to prevent any part of the body coming into contact with live parts at dangerous potentials. Subsection 412-06 refers to back-up protection by means of a sensitive RCD. This does not prevent but mitigates the effect of an electric shock by tripping the supply rapidly and usually well within the 40 ms limit specified.

Protection against indirect contact

As indirect contact shocks are the most common form of electric shock incident, five methods of avoidance are dealt with extensively in section 413.

1. Earthed equipotential bonding and automatic disconnection

Earthed equipotential bonding and automatic disconnection is the standard protection method in buildings in the UK. The relevant requirements are in subsections 413-02 and 471-08. The method consists of the creation of an equipotential zone by bonding simultaneously accessible extraneous and exposed conductive parts directly to the main earthing terminal or indirectly via the protective conductors, thus minimising the potential difference between them in the event of an earth fault. This, combined with the use of excess current protective devices such as fuses and circuit breakers, ensures that fault current is interrupted within the prescribed time. The prescribed time is matched to the risk, e.g. 0.4 s for socket outlets supplying 230 V Class I hand-held tools and 5 s for circuits supplying fixed equipment. The relaxation on disconnection times for stationary equipment in the first part of the second sentence of Regulation 413-02-09 is perhaps best implemented by a warning label and a non-standard plug and socket outlet to discourage unsafe usage.

Regulations 413-02-02 and 03 refer only to metallic parts and ignore the risk from contact with other extraneous conductive parts such as walls and floors made of, for example, brick, stone, quarry tiles or concrete which are conductive particularly when damp and which are often in ground contact and may, therefore, remain at earth potential.

Metal reinforced concrete floors could perhaps be made part of the equipotential zone by a bonding connection to the reinforcement. It should be effective if the metal mesh is electrically continuous. Where this is impracticable and there is a substantial shock risk – in wet kitchens or laundry rooms, for example – it is suggested that the location be regarded as outside the equipotential zone (see section 471-08-03) and the disconnection times reduced to those in Table 41A. Alternatively, an insulating floor covering should be provided.

When calculating the earth loop impedance, Zs, that part of it which consists of metalwork used as a protective conductor, e.g. conduit, trunking and switchboard metalwork, is normally ignored because its resistance is usually negligible. Where the earth fault loop impedance exceeds the values in Tables 41B1 and 41B2 for the disconnecting time of 0.4 s, Regulation 413-02-12 permits the time to be extended to 5 s provided the possible potential differences between touchable metalwork are minimised. This is attained by limiting the circuit protective conductor impedances to the values given in Table 41C. Where circuits requiring different interrupting times are connected to the same distribution board, precautions are required by Regulation 413-02-13 to prevent undue potential differences on touchable metalwork by utilising Table 41C or by local equipotential bonding of the distribution board and any nearby extraneous conductive parts. Table 41D gives the maximum earth fault loop impedance values for a 5 s disconnection time.

If the earth loop impedance is too high for overcurrent protection to operate within the prescribed time, Regulation 413-02-04(i) permits this time to be exceeded provided local supplementary equipotential bonding is used to meet the requirements of Regulations 413-02-27 and 28 and the fault persistence does not cause damage. Again, the intention is to minimise the potential differences between touchable metalwork items on the occurrence of an earth fault. As an alternative, Regulation 413-02-04(ii) allows the use of RCD protection.

Regulation 471-08-03 refers to locations where the installation supplies fixed equipment outside the equipotential zone, e.g. an outside motor-driven pump for a garden fountain, and the equipment has exposed conductive parts. In such cases, Table 41A applies and the maximum interruption time is 0.4 s for 230 V equipment. 230 V outdoor portable apparatus, such as an electric hedge trimmer, requires sensitive RCD protection; see Regulation 471-16-01. There are similar protection requirements for circuits in locations of enhanced shock risk such as those in Part 6; see Regulation 471-08-01.

Although the electricity supply companies have converted a substantial number of TT systems to TN-C-S systems, some TT systems are still in use, mostly in rural areas where it is often difficult or impracticable to obtain and maintain a low earth loop impedance. In these cases, the overcurrent protection should be supplemented by RCD protection as required in Regulations 413-02-18 to 20 and 471-08-06.

2. *Class II equipment*

If Class II distribution apparatus or accessories are used, Regulation 471-09-02 requires the protective conductor to be available at each wiring point to allow for the replacement of the Class II by Class I equipment. It is preferable for the apparatus and accessories to be of the all-insulated rather than the double-insulated type to avoid the possibility of exposed conductive parts making fortuitous contact with earthed metalwork.

3 and 4. Non-conductive locations

Non-conductive locations referred to in Regulations 413-04-01 to 07, subsections 413-05, 471-10, and 471-11, are for special applications only where there is skilled supervision, such as in certain types of electrical testing, as described in Chapter 17 of this book.

5. Electrical separation

The requirements for electrical separation are in subsections 413-06 and 471-12. Apart from shaver unit installations in bathrooms, the most common application of the method is in specialist applications such as some types of electrical test areas. Care must be taken in its use because of the fact that an earth fault, in the separated system, could occur and remain undetected, thus reducing the safety of the system to that of a TN system. It is therefore important that separated systems are routinely tested for earth faults, or a monitoring system is incorporated to detect such faults.

Chapter 42 Protection against thermal effects

The thermal effects associated with fire, burns and overheating are dealt with in this chapter. The treatment is quite brief because if the Regulations in the other chapters are observed, these risks should be minimal. Regulation 422-01-06 requires fixed equipment which focuses or concentrates heat to be sufficiently far away from any other fixed object to avoid damage to it. This would apply to radiant electric fires and fan heaters, for example. Regulation 422-01-07 requires enclosure materials of equipment to be able to withstand the heat generated by the equipment. Regulation 422-01-05 is applicable to oil-filled switchgear, controlgear and transformers where precautions are required to contain an oil leakage or fire. In Table 42A there are maximum touch temperatures to avoid burn injuries.

Chapter 43 Protection against overcurrent

Chapter 43 should be considered with the applications in section 473. The intention is to use automatic interruption of the supply to protect live conductors of the installation against overheating from overload and short circuit currents and against mechanical damage from electromagnetic stress. This entails the provision of fuses or circuit breakers fitted with overload trips and adequate supports for the conductors. The protection is not intended for connected equipment and its wiring, which should generally have its own protection, e.g. the fuse in a BS 1363 plug. Section 432 states that the devices must operate safely and may be for overload or short circuit protection or both. Circuit breakers, used for overload and short circuit protection or short circuit protection only, must be capable of making on to a short circuit as well as of interrupting it.

Overload protection

Section 433 is for overload protection only. It requires the characteristic of the protective device to match the current rating of the conductors so that it will operate on overcurrent before the safe temperature limit of the conductors or their insulation is exceeded.

Regulation 473-01-01 requires a device to be placed at any point where a reduction occurs in the current carrying capacity of the conductors. This often occurs where the conductor size is reduced, but it can be due to a variety of other causes which affect the conductor current rating, e.g. wiring on boilers or other locations of high ambient temperatures. It is permissible to position the device along the conductor run instead of at the current rating reduction point provided there are no branch circuits or outlets for the connection of current-using equipment between them and the installation is not in an abnormal fire or explosion risk location; see amended Regulation 473-01-02.

Regulation 473-01-03 details the cases where the omission of overload protection is allowed and in fact should not be provided because of the possible danger caused by its operation, such as generator exciter circuits and secondary circuits of current transformers. In Regulation 473-01-04 there is a list of cases where overload protection is unnecessary. In Regulation 433-01-01 there is a requirement to avoid small, long duration overloads. Thus there should be a sufficient number of socket outlets to discourage the use of adapters which could result in overloading the socket outlet receptacle contacts, and ring main socket outlet spur circuits should be fused.

Short circuit protection

Rapid disconnection of circuits subject to short circuit current is essential, although there are similar provisions for the omission of fault current protection as described above for overload protection.

The prospective short circuit currents have to be determined at all relevant locations in the installation by calculation or measurement and protective devices selected to protect all conductors against thermal and mechanical effects. For new installations the designer will first have to ascertain the loop impedances and characteristics of the excess current protection at the intake from the supply company (see Regulation 313-01-01), except where private generation is proposed. In large installations with their own substation no problem should arise, but where the supply is to be taken from the supply company's low voltage network, they will probably only be able to provide anticipated maximum and minimum loop impedance figures.

To find the loop impedance at any point the designer has to calculate the value from the intake to that point and then add the external loop impedance in order to calculate the prospective short circuit current. For small wiring, up to 35 mm^2, the inductance may be ignored so the loop impedance is the external impedance plus the resistance of the internal wiring, but for larger sized conductors the cable impedances, obtainable from the makers, should be used. A check of the characteristics of the protective devices which are in series in the circuit will indicate whether or not they are suitable for clearing the fault. For example, a motor starter may be capable of breaking the overload current of the motor but not of clearing a short circuit in the wiring between it and the motor. In this case, protection by suitable HBC (high breaking capacity) fuses may be provided so that the energy let-through of the fuses and starter will be limited to protect the starter from serious damage; see Regulations 434-03-01 and 435-01-01 and British Standard BS EN 60947-4-1 which is the standard for LV a.c. motor starters.

The standard requires the back-up protective device to operate only on a short circuit and to be coordinated with the starter, so that on the occurrence of a short circuit fault on the load side of the starter it is cleared either without, or with limited, damage to the starter.

Regulation 434-03-02 indicates that where overload protective devices are also capable of breaking the prospective short circuit current, the conductors on the load side are adequately protected. The second paragraph of this Regulation, however, indicates that this may not be valid for conductors in parallel and for certain types of circuit breaker where the excess current might cause damage to the conductor and/or its insulation if it persists. Indeed, Amendment 4 introduced a new Regulation 473-02-5 dealing with fault current protection of conductors connected in parallel. It allows for a single fuse or circuit breaker to be used to protect the cables so long as it will provide efficient protection for a single fault at the most onerous position in one of the parallel conductors, taking account of the way in which fault current will be shared and distributed. Where this cannot be guaranteed, the available options are to reduce the risk of insulation failures by mechanical protection and other means; and to provide fault current protection in each of the conductors, at the supply end when two conductors are paralleled and at both the supply and load ends when three or more conductors are paralleled.

The calculated impedance of the wiring is usually based on the full load operating temperature of the conductor, i.e. the specified ambient temperature plus the temperature rise due to the I^2R loss. The conductor resistance is proportional to its temperature so a fault current exceeding the full load current will increase the resistance and thus the impedance and diminish the fault current. BS 7454 : 1991 Method for calculation of thermally permissible short circuit current taking into account non-adiabatic heating effects, indicates how to calculate the fault current. A less accurate method is shown in Regulation 434-03-03. It is not normally necessary to do this calculation where a fault current causes rapid operation of the protection in, say, not more than 0.1 s, but where there is a significant time delay the calculation should be done.

To be satisfactory, the energy withstand of the cable K^2S^2 must not be less than the energy let-through of the protective device I^2t. If it is less, the conductor size should be increased, a heat resisting cable used or the proposed device changed.

Single phasing and neutral overcurrent protection

Subsection 473-03 caters for protection against single phasing and neutral overcurrent, including those situations where the neutral may carry more current than the phase currents in polyphase systems as a result of harmonic currents. This recognises the increasing problem associated with harmonics created by electronic switching devices such as variable speed drives and switch mode power supplies. As it is not permissible to disconnect the neutral on its own, the overcurrent device has to disconnect either the phase(s) or the phase(s) and neutral together.

A new Regulation 473-03-03 provides for the use of RCD rather than overcurrent protection in IT systems in which the neutral is not distributed.

Chapter 44 Protection against overvoltage

This is a new chapter, introduced by Amendment 4, which deals with the protection of installations against 'overvoltages of atmospheric origin or due to switching' – which is mainly concerned with the effects of lightning. The main provisions have already been described but, essentially, they require that where thunderstorm activity is expected to exceed 25 days per year, additional protection must be provided against high voltage lightning transients. In most places in the UK, this level of thunderstorm activity does not occur, so the surge protection is not needed; it can therefore be omitted, but only if the equipment connected in the installation has impulse withstand characteristics set out in Table 44A. Most equipment designed and built to harmonised standards will achieve the specified levels of withstand voltage.

Chapter 45 Undervoltage protection

Chapter 45 requires protection from any danger which may arise from a reduction in the supply voltage, either partial or total, as when there is a supply failure. This usually means that motors must not restart automatically when the voltage is restored if this would entail danger (see also Regulation 552-01-03). Motor starters should, therefore, have automatic no-volt protection which operates on low or no volts by disconnecting the supply from the motor. If damage, rather than danger, could arise from voltage reduction, the designer has the choice of tolerating it or avoiding it by, for example, the provision of standby no-break facilities.

Isolation and switching

Chapter 46 and section 476 cater for protection by non-automatic isolation and switching. There are three safety requirements:

(1)   isolating the whole or parts of the installation or equipment when not in use, as a fire precaution or to enable work on it;
(2)   switching off for mechanical maintenance on electrically powered apparatus;
(3)   emergency switching off to remove a hazard.

The nature of the devices that can be used to satisfy these requirements is described in section 537.

For the first requirement (1), the isolator has to be capable of carrying the circuit load and fault currents, but is not required to interrupt them when operated. Therefore, isolation may be effected, off load, by the removal of links, fuse links and plugs from socket outlets as well as by opening switches and circuit breakers. An important consideration for isolation is that means must be provided for securing the point of isolation to prevent inadvertent re-energisation. This is most commonly achieved by the use of padlocks applied to the likes of isolators and switched fuses; it does, however, create difficulty when, for example, miniature circuit breakers installed in non-lockable panels are used for isolation. In this latter case, measures such as applying warning/caution notices and/or locking the access door to the area in which the panel is installed may be acceptable alternatives.

For (2), uninstructed persons are likely to effect the switching and may perhaps operate the switch when the apparatus is on load, so switches and circuit breakers provided for mechanical maintenance purposes have to be of the load breaking type. Except where the switching means are under the continuous control of the person doing the maintenance, compliance with

Regulation 462-01-03 requires isolating switches to be interlocked or locked off, stop push-buttons latched off and a warning notice displayed so that the circuit cannot be unintentionally re-energised.

For emergency switching off (3), the switch or circuit breaker must be capable of interrupting any load or overload current including, for example, the stalled current of an induction motor. Circuit breakers may be manually operated or opened by means of stop push-button(s) or switch(es). Regulations 476-03-05 to 07 cover the provision of firemen's switches to isolate HV discharge lighting installations from the supply. It is not clear why such installations should be the only ones requiring such a provision. To protect the firemen in the event of a fire it would be advantageous to have similar provision for the whole installation excepting the supplies to the fire protection and emergency lighting installations.

PART 5 SELECTION AND ERECTION OF EQUIPMENT

Part 5 deals with the selection and erection of equipment. The intention is to ensure fitness for purpose. All apparatus has to comply with the relevant British Standard or comparable foreign standard. If there are no standards, the specifier should satisfy himself that the equipment meets the other requirements of the sections, including the compatibility requirements of Regulation 512-05-01 and Chapter 33, and the effect of the external influences of Chapter 32 and Appendix 5. The erection requirement includes the location of equipment in positions where it is readily accessible for maintenance and safe from mechanical damage or adequately protected against it. Dissimilar metals should not be placed in contact with each other where this might cause electrolytic corrosion and a failure of earth continuity. The owners of a new installation would be well advised to insist that the contractor provides the labelling, wiring diagrams, etc. specified in section 514, before taking over the installation, as this data is needed for safe usage and to facilitate subsequent maintenance.

Chapter 52 Selection and erection of wiring systems

Chapter 52 deals with cables, conductors and wiring materials including conduit and trunking. Regulation 521-02-01 forbids the use of single-core steel-wire or tape armoured cables in a.c. circuits and the ferrous enclosure of a single a.c. conductor because the eddy currents produced from the electromagnetic field surrounding the conductor can cause overheating. Regulation 521-04-01 sets out the standards for conduits and conduit fittings; for example, Regulation 521-04-1(iv) requires non-metallic conduits and fittings to comply with BS 4607 which demands the same fire resistance as for the non-metallic trunking, ducting and fittings referred to in section 521-05-01.

The 15th edition Regulations had a restriction on the use of flexible cables for fixed wiring, but this was dropped in the 16th edition, probably because of the increased usage of flexible cables and cords for this purpose.

The internal wiring of some machine tools is an example. The cords are drawn through the machine carcass and their multiple wires are less prone to break at terminals than a single wire conductor when subjected to vibration. It is also convenient to use short flexible cable tails and a plug and socket to connect fixed electrically powered apparatus which may have to be replaced from time to time by a non-electrically qualified person. The small, motor-driven beer pumps in hotel cellars which are usually replaced on failure by a serviceman who is not an electrician, are an example. The serviceman is competent to disconnect and connect by withdrawing or inserting a plug but not to open terminal boxes and interfere with the wiring therein. Where temporary fixed wiring is installed and then recovered for use elsewhere, as on construction sites and outside broadcast lighting installations, flexible cables are preferable as they are better able to withstand the flexing from handling during installation and dismantling. Where structural movement may occur, subsection 522-12 specifies flexible wiring.

External influences

Section 522 refers to the environmental conditions classified in Appendix 5 which have to be considered when selecting a wiring system. For installations in new premises, the designer should ascertain the likely environmental conditions and locate the wiring, if possible, where it will not be adversely affected, or otherwise match the type of wiring to the adverse conditions expected. For example, it is better to avoid the high ambient temperatures in the vicinity of radiators rather than having to increase the conductor size and/or use a heat-resisting cable insulation. Hot spots in appliances and luminaires may require the addition of heat resisting insulation over the cable cores to ensure safety should the original insulation fail. Very low ambient temperatures adversely affect some insulating materials. The PVC insulation and sheaths of flexible cables and cords used to supply portable apparatus out of doors in cold weather harden and are liable to crack if flexed so should be replaced by insulating and sheathing materials able to tolerate low temperatures.

Protection is also required against wet and/or corrosive conditions, dirt and pollutants. Regulation 522-05-02 warns against electrolytic action of dissimilar metals in contact where dampness accelerates the corrosion process, and Regulation 522-05-03 refers to hazardous degradation. An example is the degrading of PVC cable insulation when in contact with certain types of plastic loft insulation materials. Wiring accessories such as connection boxes may need sealing to exclude fine dust, and where wiring is subject to accumulations of dirt and fluff, the thermal rating of the wiring may be impaired. It will certainly be affected where it is in contact with the thermal insulating materials used in cavity walls and roof spaces. See Regulation 523-04-01 and Table 52A for the derating factors.

It is not always appreciated that some cable sheathing materials are adversely affected by the ultraviolet radiation in sunlight, and to counter this effect a black compound is added; see Regulation 522-11-01. Unfortunately, this increases the absorption of infrared radiation and this solar heating can increase the cable temperature considerably, so outdoor cables should either be screened from sunlight or derated to allow for an increase in ambient temperature of about 20°C.

Regulations 522-06-01 to 522-08-06 refer to mechanical protection against impact, vibration, abrasion, sagging and strain. Although it is permissible to bury unarmoured cables directly in walls, it is desirable to use a metal conduit or capping for mechanical protection and to facilitate replacement without damaging the wall finish. For cables buried in the ground (which are frequently struck by careless excavators) Regulation 522-06-3 requires those that are not run in a duct or conduit should have earthed metal armouring or sheathing or both, or should be of insulated concentric construction. The cables must be covered with a cable marking tape or be placed underneath cable tiles – a requirement that is frequently overlooked. The cables must also be buried at a sufficient depth – usually taken to be at least 400 mm for low voltage cables and 600 mm for high voltage cables.

Current carrying capacity of conductors

This section deals with the factors that affect the temperature rise of current carrying conductors. Table 52A gives the derating factors for cable lengths of less than 0.5 m when totally surrounded by thermal insulation. Table 52B restates the maximum continuous conductor operating temperature given in the current rating tables in Appendix 4, but also gives the maximum permissible temperature consequent on a short circuit fault which the insulation will tolerate without damage. Table 52C gives the minimum conductor sizes for various applications. As before, the calculation method is dealt with in the Appendix.

A designer will select a cable type to suit the environmental conditions. The load current and type and rating of the protective device should then be determined, then the current rating of the protective device is divided by the applicable correction factors. The factors are for grouping (Table 4B1, 4B2 and 4B3), ambient temperature (Table 4C1 or 4C2), thermal insulation (Regulation 523-04-01) and, if semi-enclosed fuses are to be used, a factor of 0.725 is applied.

The next step is to ascertain the conductor size for this current from Tables 4D1A to 4L4A under the appropriate installation method column. Next, the volt drop should be calculated from Tables 4D1B to 4L4B to ensure the conductor size is adequate for a volt drop from the supply terminals to the equipment not exceeding 4% of the declared supply voltage or the lower limit specified in the British Standard for the apparatus. Greater volt drops are permissible during motor starting. Where a more accurate volt drop calculation is necessary to take account of operating temperature and power factor, the formulae are given in paragraphs 7.1 to 7.3 of Appendix 4. The supply company has a statutory duty to maintain the voltage at the intake to within +10%/−6% of the declared voltage. This variation and the volt drop from the intake should be taken into account for voltage-sensitive apparatus, and the provision of an automatic voltage regulator considered.

Finally (and if thought necessary), the cable size should be checked to ensure its insulation will not be damaged on the occurrence of a short circuit fault by applying the formula given in Regulation 434-03-03, as discussed in 'Short circuit protection'.

In polyphase circuits, Regulations 524-02-01 and 02 warn against the use of reduced size neutrals where there are unbalanced loads and/or additive harmonic currents, and for this reason Regulation 524-02-03 forbids their use in discharge lighting circuits where the harmonic content of the phase currents is greater than 10% of the fundamental current.

Electrical connections

Section 526 covers the suitability of electrical connections. Its object is to require all joints and terminations to be properly made and secured so that they can safely carry load and fault currents, withstand mechanical stress or be in fire resistant enclosures. They should preferably be accessible for inspection, testing and maintenance. Although not a requirement, it is good practice to avoid unnecessary joints in cable runs; and where conductor sizes have been increased to cater for high ambients or thermal insulation, consideration should be given to using the larger size conductor throughout to avoid joints on either side of the hot spot.

Minimising the fire risk

The requirements in section 527 are compatible with the 1991 Building Regulations and specify that the wiring system and its method of installation will not materially add to the fire risk, so the materials used should not readily ignite or propagate a flame, and joints and terminations should be in non-ignitable enclosures as already mentioned in 'Electrical connections' above. Wherever wiring passes through the structure, non-ignitable sealing is required to prevent the spread of fire. Internal sealing of ducts and conduits is also necessary. The sections do not extend to flammable locations, where additional fire safety measures are necessary as described in Chapter 15 of this book on flammable atmospheres.

Proximity to other services

Section 528 – Proximity to other services – includes protection against environmental conditions, and is largely a repetition of section 522. Subsection 528-01 deals with the precautions to be taken when installing telecommunication, fire alarm and detection, emergency lighting and extra-low voltage circuits such as bells, to ensure their safety and freedom from electromagnetic interference. The designer should also consult the referenced British Standards. Subsection 528-02 is concerned with non-electrical services which produce adverse environmental conditions, and requires wiring to be separated from these conditions or able to withstand them, and that this wiring be so installed as to allow work on either the service or wiring without damage to either. Electrical metalwork has to be either bonded to or separated from extraneous metalwork, and cables that are not part of a lift installation are not allowed in lift shafts.

Maintainability and cleaning

The last section (section 529) of Chapter 52 and Chapter 34 – Maintainability – requires the wiring system to be selected and erected to facilitate cleaning and maintenance, so easy and safe access is required. This is particularly relevant to overhead systems for lighting, cranes and busbars. It must also be possible to dismantle and reassemble wiring accessories and the covers over wiring terminations without detriment to the original protection, so damaged cover gaskets and missing screws should be replaced, and protective conductors and neutral links disconnected for testing must be reconnected. The bridging conductors used to short out RCDs for earth loop impedance testing must be removed before restoring the circuit to service.

Chapter 53 Switchgear

Chapter 53 covers switchgear for protection, isolation and switching. It is somewhat superfluous as its requirements are mainly repetitions of the requirements of Chapter 46, section 476 *et al.,* so only brief comments are merited.

Residual current devices

As regards subsection 531-02 concerning RCDs, it should be appreciated that in addition to residual current circuit breakers the protection can also consist of a separate transformer, relay and a trip coil on a circuit breaker provided for other purposes. When selecting RCDs, those that can be safely and readily shorted out or otherwise defeated by a skilled person are preferable to enable loop impedance testing to be done on completion of the installation. If the residual current device is required for supplementary protection against direct contact electric shock, it should have the operating characteristics specified in Regulation 412-06-02(ii). Most of the devices available will trip within 30 ms at operating currents of about 25 mA or more. See section 607 for installations where there is excessive earth leakage or protective conductor current.

When RCDs are in series in a circuit and operation discrimination is required, it is usually necessary to provide a greater time delay on the supply side device. Where there are electronic loads which distort the supply waveform with d.c. components, RCDs should be selected which are designed to suit. Regulation 537-02-03 forbids the use of semiconductor devices for isolation as the circuit controlled remains live in the off position.

Switching off for mechanical maintenance

The requirements are less stringent than those for electrical maintenance as usually only part of the relevant circuits has to be made dead to prevent hazardous mechanical movement, and semiconductor devices may be used.

Regulation 537-03-02 requires an externally visible contact gap or a reliably indicated off or open position. As the switching may be effected by non-electrical personnel the alternative is preferable. In complying with Regulations 462-01-03 and 537-03-03 it is advisable to be overcautious rather than undercautious to ensure that both unintentional and inadvertent reclosure will not occur. This should prevent, for example, the type of accident which can occur on process plant which is being worked on by maintenance staff when a production worker, unaware of their activities, goes to start up a plant machine.

Emergency and functional switching

The devices for emergency and functional switching are covered in subsections 537-04 and 05 and include firemen's switches. They are required to cater for both mechanical and electrical hazards. Due regard should be paid to Regulation 537-04-04 for visibility and location. Not only should emergency stop push-buttons be provided within reach of operatives at potentially hazardous positions, but also consideration should be given to the provision of others remote from these positions and available to other people to operate should they see an operator in difficulties or other cause for emergency stopping.

In devices for functional switching, subsection 537-05 permits the withdrawal of a plug from its socket outlet up to 16 A rating, or above where the contacts have been designed to break the rated current. Regulation 537-04-02, however, forbids this practice in an emergency because the emergency might be an electrical fault entailing an excess current which could not be safely interrupted by plug withdrawal. Uninstructed persons, however, are unlikely to be aware of this, and it may be prudent to install switched socket outlets with interlocked plugs in appropriate locations.

Chapter 54 Earthing arrangements and protective conductors

Chapter 54, covering earthing arrangements and protective conductors, should be considered together with Chapters 41 and 47 in respect of indirect electric shock. Section 541 and Regulations 542-01-01 to 542-02-05 are concerned with methods of earthing the consumer's protective conductors and are dependent on the type of supply system earthing that is illustrated in the Definitions, Part 2 and in Figure 10.3 of this book. The IT system is not used in the UK for public supplies and there are very few consumer TN-C installations, so only TN-S, TN-C-S and TT systems merit comment.

TN-S system

Regulation 542-01-02 refers to TN-S systems where the supply company runs a protective conductor from the distribution transformer earth into the consumer's premises and makes it available for connection to the consumer's main earth terminal. This protective conductor may consist of the metallic sheath/armour of the underground service cable or an earthing conductor connected to the protective conductor of an overhead line.

TN-C-S system

The TN-C-S type of distribution is called PME (protective multiple earthing) of the neutral because the neutral serves as a combined neutral and protective conductor (PEN) and is earthed at the transformer and at several other locations between the transformer and consumer. The underground service is usually a concentric combined neutral and earth CNE cable, i.e. the metallic sheath is the CNE conductor. Alternatively, in the case of an overhead line supply, a similar concentric service may be used between the overhead line and consumer, or the phase(s) and PEN service conductors may be separate cables. The supply company will provide a CNE terminal at its neutral link as an earthing facility, but will only permit its use if the consumer's installation is equipotentially bonded, otherwise the consumer has to provide his own earthing facility.

TT system

Regulation 542-01-04 is about TT earthing systems. They are mostly found in rural areas with overhead supply lines. The consumer is required to provide his own earth and generally an RCD for earth leakage protection. There is an ongoing programme to convert TT to TN-C-S systems and to provide a CNE terminal for the consumer's use for earthing purposes instead of his earth electrode.

Effective earth connections

Regulation 542-01-07(i) means that the connections to earth must be effective and the earth loop impedance low enough for the protective devices to operate within the prescribed time in the event of an earth fault. Regulations 542-01-08 and 542-02-03 deal with electrolysis and corrosion. The electrolysis in Regulation 542-01-08 is probably that which occurs in d.c. supply systems where earth leakage can cause corrosion from electrolysis at the positions where the current enters or leaves metallic parts. This problem occurs in d.c. traction systems and certain types of electrochemical plants, and is often associated with cathodic protection systems which impress d.c. currents in the ground.

Regulation 542-02-03 is concerned with the corrosion of earth electrodes. According to BS 7430 : 1991, the British Standard Code of Practice on earthing, most buried earth electrodes suffer from corrosion, but provided this is not excessive, the electrodes are effective. It is important, however, to avoid corrosion of the electrode terminal. The joint should not employ metals in contact which are widely separated in the electrochemical series, and moisture and damp air should be excluded from the joint by taping or painting.

Regulation 524-04-01(iv) requires the lightning protection system to be bonded to the main earthing terminal to prevent potential differences occurring within the equipotential zone in the event of a lightning surge or strike.

Protective conductors

Section 543 deals with the types, sizes and preservation of electrical continuity of protective conductors. Regulation 543-01-01 requires a mechanically unprotected conductor to have a cross sectional area of not less than $4\,mm^2$.

For TT systems, the minimum sizes of buried earthing conductors are given in Table 54A where a buried electrode is used. For other electrodes, such as a building metal frame and for TN-S systems, the size may be selected from Table 54G or calculated from the formula in section 543-01-03. For TT systems, however, it may not be possible to calculate because the earth fault loop impedance is not likely to be known until it can be measured.

Regulation 543-01-03 allows for the cross sectional area of the protective conductor to be calculated using the formula $S = (I^2t)^{\frac{1}{2}}/k$. 'S' is the cross sectional area in mm^2; 'I' is the fault current; 't' is the operating time in seconds of the disconnecting device for a current of I amps; 'k' is a factor that takes account of resistivity, temperature coefficient and heat capacity of the conductor material, and the appropriate initial and final temperatures. Alternatively, the size can readily be selected from Table 54G. There is a series of tables, 54B to 54F, which can be used to select the value of 'k' for different types of conductor.

So far as the earthing conductor is concerned, the prospective earth fault is assumed to occur on a phase conductor between the supply company's cut-out and the consumer's main switchgear, so the earth fault loop impedance consists of the external impedance Z_E, obtainable from the supply company, plus the impedance of the earthing conductor, which is generally small and can be disregarded. The prospective earth fault current to be included in the formula for S is, therefore, the supply voltage divided by Z_E. The time 't' in the formula is found from the time/current curves for the supply company's cut-out. For small installations this is likely to be a BS 1361 fuse with the characteristics shown in Appendix 3. 'k' is obtainable from Table 54B assuming an initial temperature of 30°C. These values are then used in the formula to determine the required cross-sectional area of the earthing conductor. The formula may also be used to calculate the minimum size of the circuit protective conductors, but the earth loop impedance will now consist of the external plus the internal earth loop impedances. For cable sizes not exceeding $35\,mm^2$ their inductance is small and can be ignored, and only their resistances are used in the calculation. The impedance of protective conductors consisting of conduit, trunking or the metalwork of apparatus is normally ignored because it is negligible provided the joints are properly made.

Section 543-01-03 states that the temperature rise of conductors under fault conditions should be taken into account. This temperature rise increases their resistance and thus increases the overall impedance and reduces the fault current. The fault conductor temperature is taken as half the difference between the final and initial temperatures given in Tables 54B to 54F. The resistances should be calculated for both the phase and protective conductors. The total impedance is then the sum of the external and internal impedances. The prospective short circuit current I is the single phase voltage divided by this impedance.

The difficulty about the calculation method is in determining the prospective fault current when the supply is derived from a supply company's LV network. Changes to the network and to the installation can affect the earth loop impedance and the prospective fault current. On the whole it is probably better to use Table 54G than do the calculation and employ a smaller section conductor at what may be a small cost saving, only to find later that it was a false economy.

The minimum size of main equipotential bonding conductor for PME systems is given in Table 54H and is related to the size of the supply neutral. For new PME installations, the designer should ascertain the intended size of the supply conductors from the supply company and then select the size of the bonding conductor from Table 54H.

Earth continuity

As a discontinuity in a protective conductor is not self-revealing in service, except where circulating current earth monitoring is used, subsection 543-03 lists the precautions needed to ensure the initial and continuing preservation of electrical continuity. Joints, in particular, need careful preparation and assembly. Components of switchboards are often prepainted before construction, necessitating the removal of the paint at joints to ensure bare metal-to-metal contact. Hinged and drawout sections should have flexible bonding conductors. A precaution often overlooked is the provision of insulating sleeving at terminations on the emergent bare protective conductors of small insulated and sheathed cables; see Regulation 543-03-02.

Protective bonding conductors

Section 547 covers main and supplementary bonding conductors. Their purpose is to maintain touchable metalwork in the equipotential zone at the same potential so as to avoid the possibility of electric shock to anyone touching different metalwork items at the same time. It is necessary to check that joints in metal pipes are metal-to-metal to ensure low resistance; otherwise bonding across is needed. Look out for plumbing in which both plastic and metal pipes are used and check the earth continuity of the metal pipes. The size of the main equipotential bonding conductors is given in Regulation 547-02-01.

Supplementary bonding must be provided in bathrooms even if there are satisfactory metal-to-metal joints because of the enhanced shock risk; see section 601. Consideration should be given to the provision of supplementary equipotential bonding in kitchens, sculleries and laundry rooms with conducting floors, such as quarry tiles, particularly if they are likely to be wet, again because of the enhanced shock risk. Similar locations in commercial and industrial premises should receive the same consideration.

Chapter 55 Other equipment

Section 551 – Generating sets

Section 551 was introduced by Amendment 2 and applies to low voltage installations that incorporate generating sets running either continuously or in stand-by mode for permanent and temporary installations. This would include, for example, generators supplying temporary accommodation on construction sites and diesel generator sets operating in stand-by mode in hospitals.

Regulation 551-02-02 requires the short circuit and prospective earth fault rating of the generator to be determined, as well as the ratings of other supply sources. This will allow suitably-rated protective devices to be selected. Regulation 551-02-03 covers the rated capacity of the generator and the need to have automatic load-shedding facilities to cater for circumstances where the load exceeds the supply capacity.

Regulation 551-04-04 addresses protection against indirect contact for static inverters, typically used for uninterruptable power supplies in installations where continuity of supply is crucial. Where the disconnection times of section 413-02 cannot be achieved, supplementary bonding must be used to minimise the risk of a shock between exposed metalwork. A warning is provided in Regulation 551-04-05 about the possible deleterious effects on the operation of protective devices, such as circuit breakers, of direct current generated by the static inverter or filters.

Regulation 551-06-01 covers the interlocking arrangements that must be put in place to prevent unintentional paralleling of generators with the public supply. These can include electrical, mechanical or electromechanical interlocks on the changeover switches; locks with single transferrable keys; a three-position break-before-make changeover switch; and an automatic changeover switch with a suitable interlock. Subsection 551-07 gives requirements relating to generators running in parallel with the public supply.

Section 552 – Rotating machines

Section 552, referring to motors, requires the circuit equipment including cables to be suitable for the starting, accelerating and load currents of a motor, so the designer needs to know the characteristics of the motor, its control gear and duty cycle to enable him to determine the cable sizes and characteristics of the protection needed. Where, for example, an induction motor is to be used and started direct to line, the starting current may be some seven times the full load current so the circuit fuses or excess current trips of the circuit breaker have to be suitably selected to cater for this. If the motor is to be started and stopped more than twice per hour, larger cables may be needed to prevent overheating. There are two Regulations, 552-01-04 and 05, which deal with reverse current braking and reverse rotation of a motor from loss of a phase.

Socket outlets

Regulation 553-01-05(iii) caters for plugs and sockets with no provision for a protective conductor, e.g. circuits in SELV installations and in non-conducting locations.

Regulation 553-01-06 calls for socket outlets to be mounted on surfaces at a height where the risk of mechanical damage to the socket outlet, plug and flexible cable or cord is minimised. This would seem to require that when the socket outlet is above benches, its height should be sufficient to avoid sharp bends in flexible cords or cables for bottom entry plugs on insertion into the socket outlet, and when near the floor socket outlets should be on the wall above the skirting board to avoid damage from floor cleaning and polishing machines. The Regulation, however, says nothing about floor-mounted socket outlets now prevalent in offices with platform floors. As these are vulnerable to damage, a Regulation is needed to require them to be either safely located or protected by a barrier or housing.

Regulation 553-01-07 is a plea to avoid the tripping hazard of long, flexible cables and cords trailed across the floor. As portable apparatus may be used anywhere, socket outlets need to be fairly closely spaced, and where there is much portable apparatus to be used at one location, multi-way socket outlets are preferable to the use of adapters.

There is no mention in the standard of the positioning of socket outlets near kitchen sinks – sockets are quite frequently installed directly above sinks with an obvious risk of water/steam contamination of the socket or the risk of apparatus powered from the socket falling into the sink. The NICEIC offers guidance to its members that sockets should be located at least 300 mm horizontally offset from the edges of the sinks and this seems to be good advice.

Electrode boilers

Electrode boilers are covered by subsection 554-03. As some earth leakage is inevitable, earthing is important, so the boiler shell has to be connected to the protective conductor and to the metallic armour and sheath, if any, of the incoming supply cable and also to the supply neutral. Earthing the neutral will affect the operation of an RCD as the earth leakage current will be shared by the neutral and protective conductors. Regulation 554-03-04 refers to a three-phase electrode boiler supplied directly from an HV supply. In this case there is no neutral connection, but an RCD is a requirement. The neutral does not have to be connected to the boiler shell for boilers not piped to the water supply; see section 554-03-07.

Instantaneous water heaters

Instantaneous water heaters, i.e. where the element is in contact with the water, again need additional earthing by bonding the metal water supply pipe to the main earthing terminal; see section 554-05. Supplementary equipotential bonding will also be needed if the heater is in a bathroom or shower cubicle; see section 601.

Chapter 56 Supplies for safety services

Chapter 56 and section 313-02-01 are concerned with the requirements for electricity supplies for safety services which are generally for energising fire alarms, fire fighting and emergency lighting installations, but also, for example, for gas detection installations in unattended locations where there is a flammable hazard. Some of these installations are subject to statutory requirements which the designer must observe. He should also have regard to applicable standards and codes of practice.

The paramount requirement is reliability. There have to be alternative power sources, therefore, which automatically provide a supply on mains failure. Components of assured quality are to be preferred, where available, and installation work should be carried out by persons who are competent in the field. The wiring should be protected against mechanical damage and fire. Subsequent periodic maintenance is essential for ensuring continuing reliability.

These supplies may be used to power standby supplies to other essential services such as intruder alarm, surveillance and detection systems or uninterruptable power supplies to computer installations, provided there are multiple sources and the failure of one does not jeopardise the adequacy of the safety supply; see Regulation 562-01-04.

To maintain the integrity of the supply, Regulation 561-01-03 advocates that automatic disconnection of the supply should not occur when the first earth fault occurs, but metalwork must be adequately earthed and bonded to minimise the shock hazard; see also Regulation 566-01-01. Regulation 563-01-03 permits the omission of overload but not short circuit protection where a reduction in conductor size occurs; see Regulation 473-01-01 and 552-01-02. Where there is more than one circuit, however, overload protection is likely to be required to comply with Regulation 563-01-04.

Many safety supplies are afforded at SELV, which has the advantage of avoiding the possibility of harmful shock in the event of an earth fault. Where higher voltages are employed, separated circuits, generally complying with subsection 413-06, merit consideration. In cases where the supply is provided from a BS 3535 transformer, the connection to the incoming LV mains is often arranged on the supply side of the consumer's switchboard so as to ensure that the safety supply remains on when the rest of the installation is dead. It will usually be found convenient to use HBC fuses at this position, to provide the required short circuit protection. There is an additional Regulation 566-01-02 which requires precautions to limit circulating currents in supply sources when operating in parallel.

PART 6 SPECIAL LOCATIONS

Part 6 covers special installations or locations generally where there is an enhanced risk of electric shock.

Section 601 – Locations containing a bath or shower

Section 601 imposes special requirements in bathrooms and other rooms where there is a shower. The section was altered by Amendment 3, the main provisions of which were outlined at the beginning of this chapter.

Section 602 – Swimming pools

The enhanced shock risk in swimming pools is similar to bath and shower locations so similar precautions are specified. There are three zones – A, B and C – in descending order of risk. Zone A is in the pool, chute or flume; zone B is from the poolside up to 2.5 m vertically and 2 m horizontally from the rim of the pool, and zone C is the 1.5 m zone surrounding zone B.

In zones A and B only SELV equipment at a nominal voltage of 12 V a.c. r.m.s or 30 V d.c. is permitted excepting transformer-supplied flood-lights operating at not more than 18 V. These are usually behind a glass panel in the pool wall with access for maintenance in a passage outside the pool wall where the transformers are located. Mains voltage BS EN 60309-2 socket outlets may be used in zone B if not less than 1.25 m from the zone A boundary and at least 0.3 m above floor level and protected by sensitive RCDs or by electrical separation. These sockets are usually provided to supply pool cleaning equipment and have to be IPX4 splashproof or IPX5 if water jets are used. In zone C, IPX2 drip-proof equipment is allowed for indoor pools or IPX4 splash-proof equipment for outdoor pools. Mains voltage apparatus should be protected by a sensitive RCD or by electrical separation. This latter alternative is, however, not much used in the UK.

Supplementary equipotential bonding is specified for all three zones, but the previous requirement for an equipotential bonded metal grid in solid floors in zones B and C has been amended to bonding such grids if they exist.

Section 603 – Hot air saunas

The enhanced shock risk in hot air saunas is due to reduced body resistance because the skin is wet from perspiration. It would, therefore, be desirable to avoid contact with mains voltage apparatus as far as possible. This is, however, not made clear in the requirements. Apart from the heater, other equipment should be either excluded or mounted at a high level out of reach. Apparatus has to be IP24 fingerproof and splashproof, so the heater's flexible cord should be connected to the supply by a splashproof terminal box. Above 0.5 m from floor level, apparatus has to be heat resisting. Wiring is restricted to 180°C rubber-insulated cables which must be protected against mechanical damage using the techniques listed in Regulation 413-03-01. This requirement for mechanical protection is somewhat ambiguous but may mean either sheathed cables or unsheathed cables in plastic conduit. The standard BS EN 60335-2-53 mentioned in Regulation 603-01-01 requires a double pole isolator for single-phase and a triple pole isolator for three-phase installations. The thermostat and thermal cut-out controls and contactors, if used, have each to be fitted with temperature sensors and contacts. The heater must be clear of combustible materials in accordance with the maker's instructions.

Section 604 – Construction site installations

The special requirements of section 604 are for the temporary installations used on site for any type of building and civil engineering work, including demolition but excluding installations in site offices and other accommodation where the general sections apply.

The electrical distribution assemblies have to conform to BS EN 60439-4 and, although not mentioned in this section, contractors should carry out the installation work in accordance with BS 7375 as described in Chapter 11 of this book where more information is provided.

If the supply system is PME, the supply company is precluded from providing an earthing terminal for outdoor sites because there is no equipotential zone, so the contractor has to provide his own earth. He may well then find that the earth loop impedance is too high for operation of the excess current protection, on the occurrence of an earth fault, within the specified time and he will have to resort to earth leakage protection using an RCD.

The increased risks arising from the harsh environment are recognised by reductions in the required disconnection times for TN systems, set out in Table 604A. For example, for 230 V systems, the normal 0.4 s disconnection time is reduced to 0.2 s. It may not be possible to achieve these disconnection times because of earth loop impedance restrictions, in which case RCD protection will normally need to be specified. Note also that the 50 V touch voltage value used in Section 413 is reduced to 25 V.

Section 605 – Agricultural and horticultural premises

Section 605 prescribes rigorous safety precautions to protect animals from fire and electric shock in and around farm buildings within an equipotential zone. As livestock is more susceptible to the effects of electric shock than humans, the maximum SELV has to be lower than 50 V a.c. and appropriate to the type of animal. The Regulations, however, give no figures but imply a maximum of 25 V; see sections 605-05-09 and 605-06-01. The maximum disconnecting times in Table 605A are correspondingly less than in Table 41A and so are the values for the maximum earth fault loop impedance.

Enclosures have to be IP44 or better. All mains voltage socket outlets have to be protected by a sensitive RCD for protection against direct contact; see subsection 605-03. Indirect shock protection is afforded by earthed equipotential bonding and automatic disconnection of the supply. The disconnection times may be extended to 5 s for distribution and final circuits supplying fixed equipment. Where a distribution board supplies portable apparatus, the protective conductor impedance between the board and connection to the main equipotential bonding must be low enough to restrict its volt drop to not more than 25 V on the occurrence of an earth fault, or the conductive parts of the board have to be equipotentially bonded to all extraneous conductive parts nearby; see Regulation 605-05-06.

All exposed and extraneous conductive parts accessible to livestock have to be bonded together by supplementary equipotential conductors. Where there are metal grids in conductive floors, they have to be bonded to the protective conductor. Fire precautions (subsection 605-10) call for heaters to be separated from livestock and combustible material, and for RCD protection operating at not more than 500 mA except where supplies are essential for animal welfare.

The Regulations do not cover locations outside the equipotential zone. In such locations it is suggested that it should be possible to separate livestock from mains voltage electrical equipment or use SELV apparatus operating at not more than 25 V a.c. or 60 V d.c. Where persons only are at risk, the SELV voltages may be increased to a maximum of 50 V a.c. or 120 V ripple-free d.c. Users of mains voltage portable apparatus should be protected by a sensitive RCD, and Class II apparatus is to be preferred.

Subsection 605-14 deals with electric fence controllers which should comply with BS EN 60335-2-76 this specifies that the pulse energy into a 500 ohm load should not exceed 5 Joules, subject to certain pulse width and pulse repetition frequency limitations. These devices energise the fence wire with high voltage low-energy pulses to deter animals by electric shocks but not to kill them; normally, once an animal experiences one shock from a fence it tends not to return to it. The fence and controller are liable to be moved from time to time by non-electrically qualified persons. The fence should not be located parallel to nearby overhead power lines to avoid the possibility of the fence wire(s) being charged by induction. To avoid the possibility of the fence wire(s) becoming 'live' at mains voltage, they must be securely fixed when in the vicinity of mains voltage wiring and apparatus such as overhead power lines to prevent possible contact, and the battery of a battery-powered controller must not be recharged when connected to the fence. It is also necessary to avoid possible contact between the fence installation and telephone and telegraph wires, radio aerials and protective conductors to ensure that the HV pulses are not transferred to them.

Section 606 – Restrictive conductive locations

The more usual description for section 606 is confined conductive locations. These are spaces where freedom of movement is restricted and the body is likely to be in contact with exposed and extraneous conductive parts. This section covers work inside boilers, metal ventilation ducts and tanks, for example, where extensive contact with the metalwork increases the indirect shock hazard. The risk is enhanced if these interiors are wet or so hot that the operator's clothes are soaked with perspiration. Incidentally, although not mentioned in the sections, the precautions listed are advisable in unconfined wet and hot locations.

The best precaution is not to use electrical apparatus in these types of locations, substituting electrical apparatus for pneumatic or hydraulic equipment wherever possible. However, there are occasions when there is no option but to use electrical equipment.

For protection against both direct and indirect electric shock, the specified supply system is SELV, but the limit of 25 V a.c. or 60 V d.c. has been dropped. Where a functional earth is needed, for certain instruments for example, it may be utilised provided all exposed and extraneous conductive parts are bonded together and to the protective conductor. Where there is no need for direct electric shock protection and only indirect is required, section 606-04-01(iii) allows mains voltage supplies to fixed equipment within the equipotential zone provided the exposed conductive parts are connected to the extraneous conductive parts in the location. Alternatively, Class II equipment may be used provided its enclosure is suitable for the location and it has sensitive RCD protection to trip the circuit at a residual current of 150 mA within 40 ms (or in a time of five times the residual operating current when the current is less than 30 mA). For hand-held equipment, portable tools and hand lamps, for example, there is no relaxation and they have to be supplied at SELV.

Unless it is part of the fixed installation or it is a battery or engine driven generator providing a SELV supply, the power source has to be situated outside the restricted conductive location.

It may be impracticable to supply the more powerful hand tools, such as large angle grinders, at extra-low voltage, so section 606-04-04 allows the use of higher voltage tools protected by electrical separation. As already observed, pneumatic or hydraulically powered tools are an alternative.

Section 607 – Earthing of high earth leakage current equipment

Section 607 is mainly concerned with electronic apparatus, such as information technology equipment fitted with suppressors with an earth connection, but it is equally applicable to equipment with high leakage currents through the insulation, such as electric furnaces with resistance wire heaters in ceramic insulation. The section deals with the hazards that arise from large currents flowing in the protective conductors of such installations and is mainly concerned with the maintenance of the integrity of the protective conductors. The section was substantially changed in Amendment 4.

If the protective conductor current is below 3.5 mA, no special precautions are required. If it is between 3.5 mA and 10 mA, equipment must be permanently connected or connected using an industrial-style connector complying with BS EN 60309-2. Where the current exceeds 10 mA, the equipment must be permanently connected, or connected using BS EN 60309-2 connectors subject to restrictions on the protective conductor's cross sectional area, or there must be an earth monitoring system that automatically disconnects the supply in the event of an open circuit protective conductor being detected. There are also specific requirements for maintaining the integrity of the protective conductor where the current exceeds 10 mA; see Regulation 607-02-04.

Regulation 607-05-01 covers TT systems where the protective conductor current is so high that its product with the earth electrode resistance is 25 V or greater. That being the case, the installation must be supplied through a double-wound transformer or similar.

Section 608 – Caravans and caravan parks

Division One of section 608 deals with installations in caravans and Division Two the caravan park distribution scheme. In the UK, the caravan supplies will invariably be at 230 V single-phase 50 Hz a.c. and caravans with mains voltage installations will usually have some Class I equipment, so the protection system will be earthed equipotential bonding and automatic disconnection of the supply.

Division One

Regulation 608-01-01 makes it clear that the section applies to touring caravans and not to fixtures such as mobile homes, transportable sheds, temporary premises or structures. The supply is brought into the caravan via a BS EN 60309-2 appliance inlet and connected to a double pole isolator which can be the required RCD if combined with excess current protection. Each final circuit is controlled by a double pole MCB. Wiring must not be in metallic conduits. Although not a requirement, flexible conductors and shake-proof terminals are preferable for withstanding the vibration in transit. Provision needs to be made for securing pendant luminaires if damage is to be prevented.

Regulation 608-03-04 requires extraneous conductive parts to be connected to the protective conductor to establish an equipotential zone. The special requirements include the provisions of section 601 if there is a bath or shower, so any socket outlets must be not less than 2.5 m from them. The extra-low voltage (usually 12 V) circuit cables to the rear and brake lights have to be separated from the low voltage wiring; see Regulation 608-06-04.

Division Two

In caravan parks, the distribution system should comply with the relevant parts of the Electricity Supply Regulations 1988 as well as with the Division Two requirements. Regulation 608-13-05 could perhaps have been improved by indicating that for TN-S supplies the supply company will normally provide an earthing facility, but for TN-C-S (PME) and TT supplies an earth electrode has to be installed by the occupier. The section requires the supply socket outlets to have sensitive RCD protection complying with subsection 412-06, as for the caravan installations, so that in the event of an earth fault in a caravan both RCDs should operate. To avoid more than the faulty installation supply being disconnected, grouping should be avoided and an RCD provided for each socket outlet.

Section 611 – Highway power supplies and street furniture

Section 611 reflects the best of current practice. It applies both to the distribution wiring and connected apparatus. The off-highway areas include the outdoor parts of public parks, car parks, sports fields, bus and railway stations and fairgrounds, where movement lighting, illuminated signs and other electrical apparatus is used. The doors in lamp posts and some street pillars are usually easily openable by vandals and mischievous children so Regulation 611-02-02 requires those that are accessible to be secured such that they can only be opened using a key or special tool and there must be a fixed internal barrier to screen bare live parts.

Subsection 611-03 recognises that many installations do not have isolating switches and are fed from the supply company's combined cut-out, neutral link and sealing box on the end of the service cable, so in such cases electrical maintenance is restricted to qualified persons. The restriction does not apply to relamping if an isolating switch is available. Supplies are sometimes taken from street furniture to feed temporary installations such as market stalls, Christmas decorative street lighting and small road works. Subsection 611-06 requires no impairment of the safety of the permanent installation so it must not be overloaded and the connection must be safe. The temporary installation should comply generally with section 604 for construction sites.

PART 7 – INSPECTION AND TESTING

There are detailed requirements in Part 7 for inspection and testing of an installation, with detailed guidance published in IEE Guidance Note No. 3. These are also explained later in this book (see Chapter 17).

In the past, the installation had to be completed prior to inspection and test, but section 711-01-01 recognises that it may be more convenient and efficient to adopt a more flexible approach. It therefore permits some testing to be carried out during the progress of the work; this is appropriate for those parts of the installation to be subsequently concealed in the structure and may also have some quality control advantages. It caters, also, for larger projects that are often completed in sections and for the requirement to energise particular circuits required by other contractors, such as lift installers, before final completion. Any such energised circuits should be clearly labelled and access to live parts suitably guarded to prevent danger.

The object is to check compliance with the Regulations and the safety of the installation. To this end, the whole installation, or part of it, is first inspected and then tested. The tests are carried out in a particular order to reveal faults before the installation is energised from the supply. Before connecting the supply, the supply company will usually require a certificate from the contractor stating that the first stage tests, which indicate that the installation is safe to connect, have been done. The supply company may inspect and test the installation to satisfy itself that it is safe to connect, but it is not obliged to do so as the contractor is entirely responsible for the safety of the installation. This responsibility is shared between the designer, constructor and tester, each of whom is required to sign the completion certificate detailed in Appendix 6.

Where the work entails alterations or additions to an existing installation, some inspection and testing of it is required to ensure that it can take the additional load and is in a serviceable condition. Any defects are reported on the completion certificate; see Regulation 743-01-01. The tester is also required to enter the inspection and test intervals. In IEE Guidance Note No. 3 – Inspection and testing – Table 4A gives the maximum recommended intervals between inspections, but these may need reduction to reflect the intended usage, environmental conditions and potential hazards. A risk assessment should be done in collaboration with the user, taking these factors into consideration so as to arrive at an agreed period, which should be subsequently reviewed and altered where necessary in the light of experience.

The designer is associated with the inspection and test in so far as he has to supply the tester with the relevant design details to enable the latter to proceed. There is much to be said for him participating further as it provides an opportunity for him to appraise the finished product and ensure that it fulfils the design objectives.

To avoid damaging electronic devices by the test voltage, section 713-04-04 states that the only test required is for an earth fault, so the circuit is tested to the protective earth with the phase and neutral terminals of the device connected together.

IEE GUIDANCE NOTES

The IEE Guidance Notes have not yet been revised to reflect the changes to the standard introduced by amendments 3 and 4. There is a similar introduction in each IEE Guidance Note which says that a specification for installation work should detail the design and provide sufficient information for competent persons to do the construction work and commission it. It should also say how the system should work and provide the user with an operational manual so that he can use and maintain it safely. There is a list of those who may be concerned with the preparation of the specification, ranging from the designer to the HSE. The designer and installer would be well advised to heed this advice.

The On-site Guide

The On-site Guide was the first of the guidance documents to be published. It is suitable for the self-employed electrician who undertakes contracting work and may have to design an installation as well as construct it. The guide is of limited scope and applies only to small installations not exceeding 100 A per phase, but it is a stand-alone document and details the various steps needed from inception to completion and test.

In the foreword there is a useful reminder that an operational manual has to be prepared and supplied to the customer so that he is aware of any use limitations and is instructed in the safe operation and maintenance of the installation. This is a requirement of section 6 of the Health and Safety at Work etc. Act for the protection of all workers except domestic servants in private accommodation. Anyone, however, may seek redress if they are injured or their property is damaged by a faulty installation under the Consumer Protection Act 1987, where the provision of safety instructions is a factor taken into consideration. So the manual should be provided for installations in all premises.

For simple installations the sections do not insist on circuit diagrams but allow the relevant information to be provided in a schedule of the circuits fed from the distribution board(s). A typical example is given in Appendix 7 – Completion certificate and periodic inspection report form.

In section 7, details are given of a method of selecting the cable size for conventional final circuits, without calculation, based on wiring methods 1, 3 and 4 and the type of protection used. The maximum cable lengths are given for permissible volt drop and earth loop impedances not exceeding 0.8 ohms for TN-S and 0.35 ohms for TN-C-S systems so as to ensure disconnecting times of not more than 0.4 s or 5 s depending on the type of equipment supplied. It also applies to TT systems where RCD protection is provided.

The figures are for 440/230 V supplies. For other voltages, the maximum circuit lengths are proportional to the voltage.

Paragraph 7.3.4 is a general warning against mutual harmful effects from proximity to other services, but the only examples mentioned are for a hearing aid induction loop and hot pipes. Section 9 specifies the inspection and testing required and section 10 the procedure illustrated by pictograms.

The appendices in the On-site Guide provide much of the practical information which was in the 15th edition but omitted from the 16th. Finally, there is a pictorial index of single line diagrams for the installation, bonding and earthing, special locations requiring the use of RCDs, and inspection and testing.

Guidance Note No. 1 – Selection and erection

The third edition of Guidance Note No. 1 was published in December 1999.

Section 1 – General Requirements – is about the safety demands of Chapter 13 and some of the relevant legislation. It cites, however, only the Electricity at Work Regulations and the CDM Regulations. Attention is drawn to the responsibilities for safety of the designer, erector, maintainer and user.

Section 2 – Selection and Erection of Equipment – deals with picking equipment suitable for its intended use having regard to site conditions, EMC requirements and HV transients particularly where there is IT apparatus. The equipment should be to a harmonised standard if any, or to a BS or other standard. On page 21 of the guidance note there is a specimen circuit diagram showing the information which has to be provided to the user on completion of the installation.

Protection against overcurrent and electric shock is in section 3. It compares the performances of semi-enclosed and cartridge fuses and MCBs for both overcurrent and short circuit protection. It deals with RCDs for earth leakage and electric shock protection. It concludes with Tables 10 and 11 for the sizes of main earthing, main bonding and supplementary bonding conductors.

External influences are dealt with in section 4, which should be considered in conjunction with Appendix B which tabulates the external influences and the appropriate protection against them to be found in Appendix B for the IP Code (ingress protection) and the IK Code (impact protection). The latter is relevant to the temporary installation equipment used on construction sites.

Sections 5, 6 and 7 cover the selection, sizing and installation of cables and their protection from damage. It includes the requirements for the location of mechanically unprotected cables buried in the structure, under floorboards or underground and in Table 15 reproduces the NJUG (National Joint

Utilities Group) colour coding for underground services. There is a warning to size the neutral conductor for the anticipated current where third harmonics are likely and to avoid the risk of corrosion at connections from dissimilar metals in contact. Derating of cables in thermal insulation is mentioned as is also the degradation of PVC insulation in contact with certain thermal insulation and building materials. Installers will find Appendix A invaluable in determining conduit and trunking cable capacities and Appendix G for support methods for cables, conductors and wiring systems.

Guidance Note No. 2 – Isolation and switching

The third edition of Guidance Note No. 2 was issued in May 1999. Section 1 – Statutory Requirements – lists:

- The Health & Safety at Work etc. Act 1974
- The Electricity at Work Regulations 1989
- The Supply of Machinery (Safety) Regulations 1992
- The Electrical Equipment (Safety) Regulations 1994
- The Construction (Design and Management) Regulations 1994
- The Management of Health and Safety Regulations 1992
- The Provision and Use of Work Equipment Regulations 1992

The section also discusses their effect on isolation and switching and on the responsibilities of designers, suppliers, installers and users. Some of these Regulations have now been updated, as noted elsewhere in this book.

In section 2, the four types of isolation and switching are defined as (1) isolation; (2) switching off for mechanical maintenance; (3) emergency switching; and (4) functional switching, and in sections 3 to 6 there is a detailed explanation of each. On pages 45-47, the essential isolating procedure to be followed when work is done on any part of an installation is detailed, from which the most important point is the use of a voltage detector before touching any part normally live. This safeguard ensures that the circuit(s) are not fed from another source which has not been disconnected, and are not live from an undischarged capacitor or from inductive or capacitive coupling with another circuit.

Guidance Note No. 3 – Inspection and testing

The third edition of Guidance Note No. 3 was published in 1998 and is an amplification of Part 7 of the main standard. Advice is given on the requirements for, and procedures relating to, the initial verification of installations, their periodic inspection and testing, the reference tests to be used, the instruments to be used and the precautions to be taken during their use, and the forms to be raised to record the work.

Innovations published in the later editions of the note include the requirement for a maintenance manual to satisfy section 6 of the Health and Safety at Work etc. Act 1974; inspection of those fire-sealing arrangements of wiring penetrations, at the erection stage, which will be concealed in the structure and not subsequently readily accessible; inspection periods for a range of installations; the accuracy standard for test instruments, where HV testing is required; checking the utilisation categories of switches and isolators during construction to ensure access to the label; and the testing of PELV installations.

Section 1.1.6 provides an important reminder about the need to ensure that people who carry out inspection and testing activity must be competent to do the work. Not all electricians or tradesmen have these competencies, particularly where live working is involved, so care must be exercised when deciding who will do the work.

Table 2.1.5 in Part 2 provides very useful recommendations for the frequency at which periodic inspection and testing of different types of installations should be conducted, as well as the frequency of periodic routine checks. For example, for restaurants, hotels and hospitals, the maximum interval between inspections and tests should be five years, whereas it should be one year for the likes of cinemas and caravan parks. A summary of what periodic testing should be done is included in Table 2.3 and a checklist for inspections is contained in section 1.2. The intention is that these should be the initial intervals, used as a baseline, such that the intervals may be varied according to experience. HSE's electrical inspectors will use this table as the benchmark standard for compliance with the Electricity at Work Regulations, Regulation 4(2), although they will take a discretionary approach to its application.

Section 2.2.3 External Influences is a useful reminder to look for alterations which have occurred during construction, not foreseen by the designer, and which are detrimental to the wiring or equipment. Additional items to look for are in section 4 of Guidance Note No. 4 – Protection against fire.

Part 3 contains helpful diagrams to explain how the basic low voltage installation tests should be conducted and Part 4 stipulates the requirements for the test equipment to be use.

Part 5 describes the forms that should be used to record the work. There is no legal duty on anybody to use these forms or, indeed, to record the results of the inspections and tests, although it may be required as part of licensing agreements in hotels, restaurants and so on. However, it is very good practice to do so. Recording the work provides a means of demonstrating compliance with the duty to maintain, and affords the opportunity to detect deterioration in values of, say, insulation resistance over time.

Guidance Note No. 4 – Protection against fire

The third edition of Guidance Note No. 4 was published in November 1998. Section 1 lists the statutory requirements including the 1990 Scottish and 1991 English Building Regulations. Both have electrical requirements for fire stop, and the former, in section 26 for electrical installations, calls for avoidance of injury.

Section 2 draws attention to the need for special consideration of installations where it is desirable for apparatus and wiring to be fire resistant and to be low emitters of smoke and toxic gases when affected by fire.

Section 3 – Thermal effects – identifies many of the items that are potential sources of ignition and indicates some of the appropriate precautions. More might have been said in section 3.2.3, perhaps, about possible failure modes of pinch screw connections. Section 3.2.3(ii) is not clear but presumably means that overheating may occur if the mechanical stress breaks some of the conductor wires or pulls some of them out of the termination.

Section 3.7 lists dusts which could be an explosion hazard. It refers to dust clouds ignitable by a spark. Probably the item 'spray paint' should be 'spray powder paint', as applied by electrostatic spraying, which is prone to ignition from a static spark.

Section 4 is concerned with fire risks consequent on alterations or additions. Among the electrical causations listed are overloading of cables due to changes in the connected load, cables surrounded by thermal insulation applied subsequent to the initial installation, and extra circuits added in conduit and trunking. Some of the non-electrical causations are the subsequent installation of steam pipes close to parts of the electrical installation, and changes in the use of the premises which may introduce adverse conditions inimical to the installation, such as dust, corrosive fumes or vibration.

Section 5 – Locations with increased risk – is concerned with potentially explosive atmospheres and the processing or storage of flammable materials. For the former, the precautions detailed in Chapter 15 of this book are advocated, and for the latter, separating the electrical equipment from the flammable materials as far as possible is advised and for the rest, in proximity to the material, taking special precautions to minimise the danger. Cables, with low emissions of smoke and corrosive gases, are recommended in places where people congregate or where there are escape problems.

Section 6 – Safety Services – covers emergency lighting, fire alarms and lightning protection.

Section 7 is for cable selection and section 8 for maintenance.

Guidance Note No. 5 – Protection against electric shock

Guidance Note No. 5 is an important publication as it provides guidance on the main objective of the Regulations which is to safeguard persons and animals against electric shock and its consequences. Section 1 deals with the seven Electricity at Work Regulations which are concerned with electric shock precautions. Section 2 – Protection against direct contact – lists the precautions needed, i.e. insulation, barriers and enclosures, obstacles and placing out of reach. In respect of the latter, the Guidance Note points out that placing out of reach should only be used in suitable locations, which would rule out places where long metal parts are handled. In section 2.7 there is some guidance on the use of sensitive RCDs as back-up protection, and in section 2.8 – Other measures and precautions – it is pointed out that there are few applications for protection by limitation of discharge of energy. It is mainly used for electric fences and electrostatic paint and powder spraying.

Section 3 is lengthy and deals with indirect shock protection in detail. This is justified, as indirect shocks are far more prevalent than direct shocks. The prescribed protective measures, listed in Table 2, are:

- earthed equipotential bonding and disconnection of the supply;
- use of Class II equipment;
- non-conducting location;
- earth-free local bonding;
- electrical separation.

The first of these measures is the most widely used, and its requirements are generally understood by the electrically qualified, except perhaps for the bonding of some extraneous metalwork such as metal framed windows where it is often a matter of risk assessment and judgement; see section 7.2. The relation between disconnection times and degree of risk is explained, including the 5 s 110 V centre tapped to earth (CTE) system used on construction sites.

There is only limited use of Class II equipment in the fixed installation because in most cases it has to include the protective conductor for use elsewhere in the circuit to connect Class I equipment and this usage negates the Class II classification. Class II is more prevalent in portable apparatus. Protection by non-conducting location is not much used in the UK except for special locations such as test facilities. Earth-free local bonding, again, is used only in special locations where the necessary precautions can be taken to avoid importing an earth. Electrical separation's main application is also in electrical testing areas.

Section 4 deals with yet another method – the use of extra-low voltage, i.e.

not exceeding 50 V a.c. or 120 V ripple-free d.c., to minimise the shock danger. It discusses the precautions needed and some of the applications.

Section 5 – Earthing – is about the earthing methods in the several supply systems and the requirements of the Electricity Supply Regulations. Section 5.2 deals with the problems of earth electrodes where the consumer has to provide his own earth.

Section 6 – Circuit protective conductors – cites the circumstances where an additional circuit protective conductor (CPC) is likely to be needed.

Section 7 deals with main and supplementary equipotential bonding and section 8 with the enhanced risks in the special locations of Part 6 of the sections. In Table 3 the increased risks are analysed and in Table 4 the precautions required for each location are summarised. In section 8.4 there is some useful information about equipotentially bonded floor grids.

Section 9 deals with the electric shock hazard when the supply is PME and there is a potential difference between the PME terminal and the Class I metalwork bonded to it and true earth, and the precautions needed to ensure safety. It fails, however, to mention the worst case (admittedly rare) which arises from a break in the PEN conductor.

Appendix A – Maximum permissible measured earth fault loop impedance – is a rewrite of Appendices 7 and 8, and Appendix B – Resistance of copper and aluminium conductors under fault conditions – is a repeat of Appendix 17 of the 15th edition. Appendix C – Minimum separation distances between electricity supply cables and telecommunication cables – tabulates the separation specified for external and internal cables. Appendix D – Permitted leakage currents – gives the figures culled from the relevant British Standards for fixed and portable equipment.

Guidance Note No. 6 – Protection against overcurrent

Guidance Note No. 6 was revised in 1999 and gives extensive guidance on both overload and fault current protection and the limitations of the devices used. Section 1.6 says that although the sections do not limit the duration time of overcurrents unless they cause damage, there is a requirement to so design the circuits as to make small overloads of long duration unlikely. This entails an assessment of the circuit loadings and diversity. Where the over-current is an earth fault current, the current path may cause unexpected damage elsewhere and perhaps a fire from arcing. Undetected earth leakage through a high resistance fault can also be a fire hazard because it can persist for a long time without causing the overcurrent protection to operate. The remedy is earth leakage protection.

Section 1.7 deals with discrimination and coordination and cites the protection of motor starters as an example where the back-up protection has to interrupt high fault currents before the starter protection to avoid damaging the starter. In sections 2 and 7 load assessment is further discussed together with cable sizing leading to the selection of suitable overcurrent devices. Section 3 for fault current protection includes the thermal and electromagnetic effects of such currents. Sections 4, 5 and 6 are devoted to measuring or calculating potential fault currents.

Although the sections do not extend beyond the socket outlet, section 8 is about fault current withstand of the flexible cord connecting the appliance. There are few problems where BS 1363 plugs and sockets are employed because of the local fuse protection in the plug top, but where there is no local fuse and the protection is in the distribution board, the size and length of the flexible cord which can be safely used is listed in Table 9. Appendix 1 – Calculation of reactance – applies to the larger cable sizes where the reactance is a significant part of the loop impedance. Appendix 2 provides a formula for the calculation of k for temperatures other than those given in the tables in section 543-01-03.

Guidance Note No 7 – Special locations

Guidance Note No. 7 covers the special locations addressed in Part 6 of BS 7671. It was issued in April 1998, covering BS 7671 up to and including Amendment No. 2. It therefore does not contain information on the more recent significant amendments to section 601 on bathrooms and showers and section 607 on installations having high protective conductor currents.

Chapter 1, bathrooms and showers, contains a helpful explanation of why there is an increased shock risk in wet locations and what needs to be done to protect people against those enhanced risks.

Chapter 2, on swimming pools and fountains, contains at Figures 2.1 and 2.2 a diagrammatic representation of the zones prescribed by the standard. Chapter 3 briefly covers sauna installations.

Chapter 4, on construction sites, provides a comprehensive explanation of reduced low voltage systems and, at Table 4A, gives recommendations on the frequency of inspections and tests for installations and equipment used on construction sites.

Chapters 5, 6, 7 and 8 cover sections 605, 606, 607 and 608 respectively, although Chapter 7 is now slightly outdated as a result of the changes brought in by Amendment 4.

Chapter 9 introduces new guidance on electrical safety in marinas, based on IEC and draft CENELEC standards. There is no matching section in BS 7671. The guidance is based on the fact that marina locations have increased risks of electrical injury arising from the accelerated deterioration caused by movement of the boats and the wet and salty nature of the location. The general requirements therefore stipulate that the precautions should take account of increased corrosion, movement, mechanical damage, and reduced body impedance. Supplies to boats must not be PME, for the same reason that supplies to caravans must not be PME; either TT or TN-S systems are specified, with suitable RCD protection. Detailed recommendations are made for wiring systems, distribution boards and socket outlets.

Chapter 10 covers electrical installations in medical locations and associated areas. As in the case of Chapter 9, there are no equivalent regulations in the main standard and the guidance is based on IEC and CENELEC standards. The chapter considers the use of electrical equipment in three main areas: life-support equipment (such as infusion pumps and dialysis machines), diagnostic equipment (such as X-ray machines and blood pressure monitors) and treatment (such as defribillators). It comments that the risks of electrical injury are enhanced by factors such as the acute nature of the care that patients may be undergoing and the fact that some treatments may be given when the skin barrier is broken, leading to reduced total body resistance.

The chapter divides hospital locations into three groups, according to the nature of the treatment being provided. Group 0 locations are where no low voltage medical electrical equipment is being used; Group 1 is for locations where such equipment is intended to be used but not for procedures which involve invasive use of conductors near to the heart (so-called intracardiac procedures); and Group 2 locations are where medical electrical equipment is intended for use in intracardiac procedures. Table 10A allocates typical medical locations to an appropriate group.

Section 10.4 provides recommendations on the type of supply to be used for the groups, stipulating, for example, that SELV supplies must be limited to 25 V r.m.s. a.c., and that live parts must be protected against direct contact.

Section 10.5 provides extensive advice on the precautions to be taken against indirect contact, including a recommendation that separated systems with insulation monitoring facilities should be used for life-support equipment. This is to prevent such equipment being automatically disconnected in the event of an earth fault, which would be an undesirable outcome.

Section 10.7 covers control of the explosion risk associated with the flammable gases used in medical treatment areas.

Section 10.8 summarises the requirements for standby power supplies for safety and emergency services, grouped depending on the required change-over time. This means that consideration needs to be given to the provision of uninterruptable supplies in some circumstances, and stand-by supplies in others.

Section 10.9 makes reference to the Department of Health's technical memoranda that cover electrical system requirements in medical establishments.

Chapter 11, on highway power supplies and street furniture, reflects the requirements of Chapter 611 of BS 7671, explaining that the requirements relate not just to street furniture on highways but also to installations in places such as private car parks and private roads. The term 'street furniture' includes the likes of road lighting columns, traffic signs, bus shelters and advertising signs.

One of the main external influences for this type of equipment is vandalism, and the precautions against injury as a result of vandalism are covered. This is mainly in terms of providing internal IP2X equipment or barriers to prevent finger contact when the furniture door is opened, and ensuring that access doors are resistant to vandalism.

Section 11.3 indicates that it is acceptable to use the fuse carrier in TN systems as the isolation and switching device, subject to a limiting current of 16 A. A number of incidents occur when street lighting personnel insert fuses on to circuits with short circuit faults, leading to flashover burn injuries. It is a wise precaution for circuits that have been subjected to maintenance activity or possible physical damage to be tested to ensure that they are fault-free before inserting the fuse.

Section 11.6 covers temporary supplies taken from street furniture and expresses concern about the potential for damage to the permanent wiring. It makes the sensible recommendation that socket outlets should be installed inside the compartment to supply such loads, subject to the temporary loads not being the cause of overloading.

Section 11.7 recommends that temporary decorative loads such as Christmas tree lights should be supplied at SELV. In some cases, this compromises the attractiveness of the decorations and there is a temptation to supply them at 240 V to achieve the desired brilliance and luminance. In that case, if the lighting fixtures can be accessed by the public, there will be a need to provide RCD protection to minimise the risk of electric shock injuries, although the potential problems arising from nuisance tripping will need to be considered.

Chapter 12, on exhibitions, shows and stands, is another chapter that does not have an associated section in Part 6 of the standard. It covers internal and external installations where there are particular risks arising from their temporary nature, the mechanical stresses involved, and the fact that the general public has access to them.

PME supplies are not considered to be acceptable, with a TN-S supply being acceptable if one can be made available. Failing that, TT supplies will be most common. However, the disconnection times are lowered for the TN systems, with consequentially lower earth loop impedance values. All final circuits must have 30 mA RCD protection, with 300 mA to 500 mA RCDs installed on distribution circuits subject to increased risk of physical damage. The increased risk of fire, associated with installations in marquees for example, is addressed in section 12.5 The increased risk of mechanical damage to wiring systems and enclosures is covered in section 12.9. Section 12.10 covers the provision of supplies from generators – a frequent occurrence for outdoor events. It recommends that TN or TT supplies should be referenced to earth using earth electrodes. This, of course, can sometimes be difficult to achieve when the event is being held in locations such as the middle of a hard standing – a large car park, for example, with no street furniture which could be used as an earth electrode. The section could usefully be expanded to provide advice on how to tackle this type of 'real-world' problem. In fact, most low voltage generators used in these types of events can be configured as TN systems, with the neutral conductor or star point of the generator connected to the frame of the generator, which in turn is connected to the exposed conductive parts of the installation via a protective conductor. Whereas it is preferable for this type of temporary installation to be earth-referenced, and they should be wherever practicable, it is not absolutely essential so long as all other requirements relating to the safety of the installation are followed.

The increasing use of electrically-supplied features in gardens, such as fountains in ponds and garden lighting, makes Chapter 13 – Gardens (other than horticultural installations) – an important addition to the guidance material. There is no equivalent chapter in the standard, although the guidance in section 605 is obviously pertinent. Advice is provided on safety associated with buried cables, supplies taken outside from socket outlets in the equipotential zone of the property, decorative lighting and electrical equipment in ponds.

Chapter 11
Construction Sites

INTRODUCTION

The construction industry has an unenviable accident record and, until the 1990s, had more electrical fatalities than any other industry. The accident rate dropped during the 1990s, largely as a result of the emphasis placed on the use of 110 V centre-tapped-to-earth supplies to hand tools and the use of safer distribution systems. However, serious accidents and fatalities still occur, with contact with overhead power lines being an all too frequent occurrence.

The accidents tend to dominate on building sites where there is a temporary electrical distribution system rather than at works of engineering construction such as road building where little or no electrical apparatus is used. There is, however, a significant number of burn accidents in street works when underground electricity supply cables are damaged during excavation activity. This is, perhaps, not surprising given that the number of cable strike incidents is in the order of 50,000 per year.

From an historical perspective, the problem with electrical safety on construction sites largely began after World War II when portable electric tools, electrically powered plant and electric lighting came into general site use. Up until the 1960s, there was no purpose-designed electrical distribution system available, so the main contractor would usually ask the electrical subcontractor to provide a minimum installation at minimal cost.

The usual practice was to erect a few switch fuses, distribution boards, socket outlets and lighting points on an ad hoc basis, connect them with unprotected PVC-insulated conductors draped over the structure and operate the installation at mains voltage. Non-weatherproof apparatus, of the type used for indoor fixed wiring installations, was employed – usually with indifferent quality to save cost and often with insufficient robustness for site use.

As the work progressed, the temporary installation had to be altered to suit, involving repositioning apparatus and wiring. This was done with scant regard for safety. The apparatus and wiring were seldom effectively secured, were vulnerable, and sustained a great deal of damage. At the end of the contract very little of the temporary installation was serviceable and reco-verable for use elsewhere, so it was not surprising that most site installations were illegal, cheap and nasty and a recipe for electrical accidents.

To counter the rising accident toll, the Building Research Establishment designed and built a safe site distribution system which came into use in the early 1960s when several manufacturers began to produce the equipment. The system was accepted in 1965 by the National Joint Council of the Building Industry who recommended its progressive adoption, over a five-year period, with a final implementation date of 1 January 1970. This system is now accepted as the benchmark for safe electrical systems on construction sites and is reflected in existing standards and guidance material.

SAFETY CONCEPT

The safety philosophy is based on dividing the site installation into two distribution systems: the 400/230 V system where there is a comparatively high shock risk because the voltage to earth is 230 V, and the comparatively safe 110 V system where the voltage to earth does not exceed 64 V. The 400/230 V system is intended to be installed and maintained only by authorised and competent persons, i.e. those who are electrically qualified. The con-struction workers' role is confined to operating the distribution equipment, which is metalclad, weatherproof and earthed. The portable apparatus which they constantly handle is all connected to the 110 V system (see Fig. 11.1).

THE STANDARDS

BS 4363 Distribution units for electricity supplies for construction and building sites, was issued in 1968, and was amended in 1992 for compatibility with BS EN 60439-4 : 1994. This standard specifies the equipment. In 1969 BS CP 1017 appeared. This was the code of practice which describes how the BS 4363 apparatus was to be installed and used. This code was subsequently amended and renumbered as BS 7375 : 1996. See also HSE guidance booklet HS(G)141 – *Electric Safety on Construction Sites*.

The distribution equipment comprises a number of metalclad, fireproof and weatherproof units with airbreak switches and circuit breakers. Although not a requirement in the standard, the transformers used in practice are generally of air-insulated or resin-encapsulated construction. The apparatus is robust to withstand rough handling, free-standing and

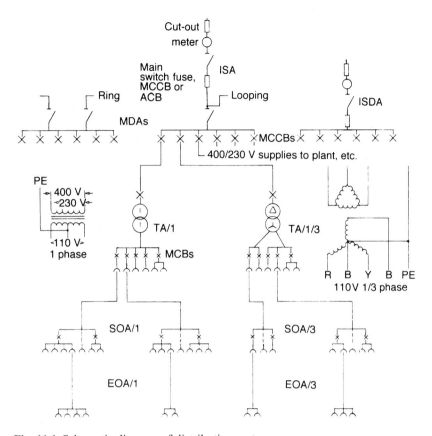

Fig. 11.1 Schematic diagram of distribution system.

mounted on ground clearance supports with facilities for anchoring. It is totally enclosed, and unauthorised access is prevented by bolted-on or lockable covers. To facilitate handling, the heavier units are provided with lifting lugs and the lighter with handles.

Incoming supply assembly

To cater for the supply company's needs, the standard describes the requirements for an incoming supply assembly (ISA). It has two compartments with lockable access – one for the incoming cable termination, service fuses, neutral link, current transformers and meters, and another for the consumer's main switchfuse or circuit breaker fitted with excess current protection and, where appropriate, earth leakage protection.

Main distribution assembly

The main switchboard is called a Main Distribution Assembly (MDA) and consists of an 'on load' isolating switch which feeds a set of bus-bars to which are connected a number of moulded case circuit breakers (MCCBs) (see Fig. 11.2). For smaller sites and/or where it may be convenient, the ISA and MDA can he joined to form a combined incoming supply and main distribution assembly (ISDA). In this case the 'on load' isolator is omitted.

Fig. 11.2 Blakley Electrics 400 A MDA with on-load isolating switch, incoming MCB with overload and adjustable earth leakage protection and seven outgoing MCBs with sensitive RCD protection (front and side views).

Fig. 11.2 (continued).

Transformer assembly

For supplies to portable electric tools, small plant items and the part of the lighting installation requiring a 110 V supply, transformer assemblies (TAs) are used. The transformer is controlled by means of an incoming MCB with excess current protection (see Fig. 11.3). The secondary output circuit feeds a number of BS EN 60309-2 : 1992 socket outlets controlled by MCBs with excess current protection. Single-phase and three-phase TAs are available. The secondary winding has its centre point for single-phase, or its star point for three-phase units, earthed to limit the potential electric shock value to earth to 55 V or approximately 64 V respectively. Single-phase 110 V supplies may be obtained from three-phase units by utilising two phases.

For jobbing work, small, single-phase, portable TAs are available up to 2 kVA capacity. A length of flexible cable is provided, usually terminating in a BS 1363 fused plug for insertion into 230 V domestic socket outlets. The

Fig. 11.3 Blakley Electrics 10 kVA 230/110 V TA with incoming MCB and four 16 A and two 32 A socket outlets protected by MCBS.

output circuit is controlled by an MCB with excess current protection feeding two BS EN 60309-2 : 1992 16 A socket outlets.

Socket outlet and extension outlet assemblies

To improve the versatility of the TAs and facilitate the rapid provision of 110 V supplies anywhere on site, multiple socket outlet assemblies (SOAs) and extension outlet assemblies (EOAs) are specified. An outlet assembly has an incoming length of flexible cable terminating in a BS EN 60309-2 plug for insertion into a 110 V socket outlet on a TA. The cable is connected to busbars feeding a number of BS EN 60309-2 socket outlets controlled by

MCBs with excess current protection. The EOAs are similar but without the MCBs. The EOAs can be fed from the socket outlets on either the TAs or SOAs. The SOAs and EOAs are light enough to be carried by one man.

INSTALLATION PRACTICE

Initial planning

The Construction (Design and Management) (CDM) Regulations require the installation designer to cooperate with other designers, such as the architect, the planning supervisor and client, to ensure that the installation will be safe both initially and subsequently as it is altered to suit the various phases of the construction work. This entails a consideration of the risks and a determination of the means to be used to minimise them. When the site work begins, the electrical contractor responsible for the temporary installation has to plan his work so that it can be done safely.

At the initial planning stage and as soon as the maximum demand has been assessed, negotiations with the electricity supply company should start so as to ensure the supply will be available when required. In some cases the electricity supply company may have to obtain wayleaves for an overhead line to supply the site and this can take months. For the larger sites, with a substantial demand, the supply company may have to supply at high voltage and will then require a site location for a temporary substation. When it is known that a permanent substation will be required to supply the development, it will sometimes be economic to build and equip the substation first and use it to supply the site.

The position of the supply intake where the ISA is situated should be agreed with the electricity supply company and should be located where it will not be disturbed for the duration of the project and where it is not liable to be damaged by vehicles and mobile civil engineering plant. The location will usually be on the site periphery and should be selected to minimise the length of the supply company's service cable or overhead line and consequently its charges.

If work has to commence before a public supply is available, a temporary supply may be obtained from a mobile, engine-driven generating set which could be hired.

The electricity supply companies generally use single-core PVC-insulated and sheathed conductors without further protection for the connections between their cut-outs, metering equipment and the consumer's main switchfuse or circuit breaker, and very often this wiring is not properly secured. It is not, therefore, as safe as the site installation wiring, so access to it should be restricted to authorised and competent persons. Any building or hut in which the equipment is installed should not be used for other purposes, such as an electrician's store, and the door should bear a danger notice.

The MDA is akin to a main switchboard and should be located at the load centre. On scattered and on large sites it may be convenient and more economical in subcircuit cabling to employ several MDAs at local load centres. The locations chosen should, preferably, be in positions where the equipment can remain for the duration of the contract as relocating could prove inconvenient, time-consuming and expensive. To facilitate connecting the MDAs on a ring main, or several on a single cable, double incoming terminals should be provided for looping.

The location of the TAs is not so critical as for the MDAs as they are more readily disconnected and can conveniently be moved from time to time, to suit the needs of the users of the 110 V portable tools and apparatus.

Distribution cables

BS 7375 specifies armoured cables for site distribution, except where the risk of mechanical damage is slight, and recommends plastics-insulated armoured cables. There are not likely to be many locations free from the risk of mechanical damage so it is probably wise to standardise on PVC-insulated, single-wire-armoured and PVC-sheathed cable to BS 6346 and to equip the distribution units with glands and armour clamps suitable for this cable. Screwed spout entry, bolted-on glands are better than securing the gland with lock nuts on each side of a hole in the gland plate. The former method provides a joint that is less likely to corrode and/or loosen. Glands with earthing terminals are available and are recommended as they enable a direct bond to the earthing terminal or busbar to be made.

There are differing views on installing the cables. On balance, it is probably better to run the cable on the surface where it is visible, rather than bury it. Buried cable routes are supposed to be surface marked, but on construction sites it is difficult to preserve the markers and the cables then become vulnerable to damage from excavators. They are more difficult to recover on project completion for use elsewhere, whereas surface cables are easily recovered. Although surface cables are more readily damaged, they are tough and resistant to mild abuse. Where they cross site roads, however, damage should be prevented by installing them in buried ducts or in steel pipes within surface ramps.

Cables subject to frequent movement, such as those used to supply mobile cranes, should be flexible to BS 6708. This standard calls for a protective conductor other than the armour and for circulating current earth monitoring to prove its integrity (see Fig. 3.1). Cables in the 110 V system are usually unarmoured plastics-insulated and plastics-sheathed. The safety feature is the comparatively low voltage to earth so damage does not cause a serious shock hazard.

Overhead lines

The code of practice discourages the use of overhead lines for site distribution at mains voltage because of the shock hazard from contact with the jibs of cranes and excavators, bodies of tipping lorries, ladders, scaffold poles, etc. It is not easy to avoid such contacts, which are responsible for a number of electrocutions and electric burn fatalities annually. The Electricity at Work Regulations 1989, Regulation 14(c), stipulate that when work is being carried out adjacent to live overhead lines, suitable precautions must be taken against injury, as spelled out in Guidance Note GS6.

An overhead line distribution system is unlikely to be economic as any saving in initial cost is likely to be more than lost by the cost of fencing off the route to comply with the law. Furthermore, an overhead system and its fence is not readily rerouted from time to time to suit the progress of the work. Overhead line systems are, therefore, best avoided.

If there is a supply company's overhead line traversing the site, and it is low enough to be a hazard, the distribution company should be asked to divert it at the initial planning stage. Not only is this desirable for the duration of the project but it may be necessary anyway to cater for the subsequent use of the site. If the line cannot be diverted, the company should be asked to make it dead for the job duration. If it is impossible to do either, the line will have to be safeguarded. Figure 11.4 shows how this should be done. If the line runs along the site boundary it is only necessary to provide barriers on the side where the plant is in use. If plant has to cross under the line, site roads will be required at each crossing point and should be situated so that there is adequate vertical clearance of not less than 5.2 m where there is no motor transport or mobile plant, and 5.8 m where there is. The road surface should not be so uneven that the lowered plant jibs are tilted upwards, nor so rough that they bounce and hit the line conductors. Barrier metalwork which could become live either from inductive charging, e.g. a wood post and wire fence, or from contact with the overhead conductors, e.g. a wire crossbar to wide goalposts which might be caught by a jib and forced into contact with a line conductor, should be earthed.

Warning notices should be provided to instruct drivers to lower jibs and advise the height of the crossbar. The barriers should be situated at not less than 11 jib lengths from the line to ensure adequate clearance. There is no specification for the barriers which may be of the type that would stop the plant, e.g. ditches, earth mounds, timber or concrete balks, or of the type that would not, e.g. wire or rail fences and rubble-filled drums. If the latter type is

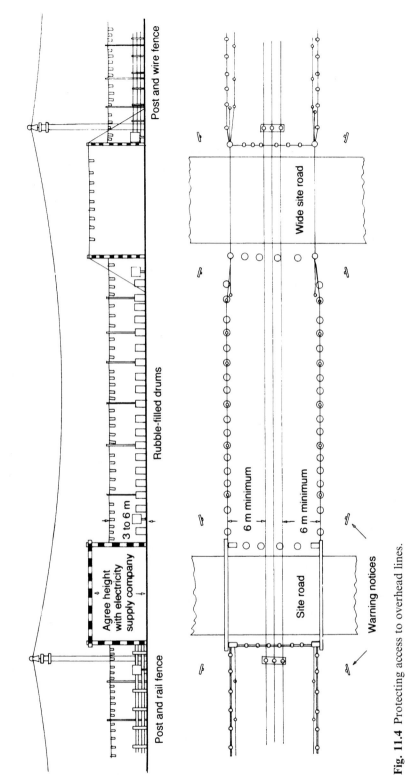

Fig. 11.4 Protecting access to overhead lines.

used, bunting, at a higher level, should be displayed to attract the driver's attention. Proximity warning devices are not acceptable as a substitute for barriers but may be used as an additional warning device. For night work, the location of the barriers should be indicated by suitable lighting and the warning notices should be illuminated.

Barriers are not always appropriate for cross-country pipeline laying and the like, where the work site is constantly on the move. For the comparatively short time when the work is being carried out beneath the line, it is sometimes possible to arrange with the line owner for it to be made dead. If this cannot be done, a jib restraining device should be fitted to prevent the jibs of cranes, excavators, etc. being raised to a dangerous height. It is not usually economically practicable to erect a temporary insulating or earthed screen below the line to prevent jib contact, and the alternative of relying on a banksman to control the driver is not entirely satisfactory as it is subject to the human error of both persons.

Electrical protection

Short circuit rating

Protection against overcurrent and short circuit current is required and is mainly provided by circuit breakers fitted with overload trips. At the initial planning stage the supply company should be asked for the prospective fault level at the supply intake so that adequately rated equipment may be selected. At this position, if a switchfuse is used to control the installation and the BS 88 HBC fuses specified in BS EN 60439-4 are employed, there should be no problem as the breaking capacity is not less than 80 kA at 400 V. If a moulded case circuit breaker is used, however, it is necessary to select one of adequate fault breaking capacity as they are made for a range of ratings.

If the supply company provides a supply from its LV network, the quoted fault level is unlikely to exceed 16 kA, but on large sites where a substation is needed, a higher fault level could occur; about 27 kA for example, at the output terminals of a 1000 kVA transformer with 5% impedance. The prospective fault level becomes progressively less as the distance from the intake increases, mainly due to the impedance of the distribution cables, so a lower fault rating for the MDAs and MCCBs is permissible.

At the initial planning stage the prospective fault level should be calculated for several locations in the 400 V distribution system so that apparatus of adequate fault rating may be used. Inadequately rated apparatus, for instance an automatic motor starter, can often be protected by back-up HBC fuses, providing they have a suitable time/current characteristic so that they will rupture before the contactor is damaged by arcing in the event of a short circuit.

Electric shock – interrupting times

Protection against electric shock at mains voltage and from indirect contact consequent on an earth fault, i.e. contact with live metal other than the conductors, may be provided by means of the excess current protection. It will be satisfactory if the earth loop impedance is sufficiently low for enough earth fault current to flow to operate the protective device within the recommended time. BS 7671, section 604 for construction site installations, recommends in Table 604A a maximum disconnecting time of 0.2 s for 230 V and 0.05 s for 400 V manually handled moveable equipment such as floor grinders and cutters connected by means of an armoured flexible trailing cable. For mains voltage fixed equipment and large mobile plant (items such as electrically powered cranes) the maximum disconnecting time is 5 s.

To obtain an interrupting time of 50 ms, it will usually be necessary to use an RCD. The earth fault current I_F is the phase voltage, normally 230 V, divided by the earth loop impedance U_o/Z_S. The disconnecting time, t, for the earth fault current I_F is found from the time/current characteristics for the relevant fuse or circuit breaker, obtainable from the maker. However, in BS 7671, section 604, the maximum earth fault loop impedances are given for disconnection times of 0.2 s for a range of fuses and MCBs, and the time/current curves are in Appendix 3.

BS EN 60439-4 recommends that all portable apparatus should operate on the comparatively safe 110 V system where the tripping time, for shock protection, is not important as the safeguard is the low voltage to earth, and BS 7671, section 604 allows a disconnection time of 5 s. If, however, 230 V portable apparatus is used, BS 7671, subsection 471-16 recommends that electric shock protection should be provided by means of a sensitive RCD with a rated operating current of not more than 30 mA. Such devices also afford protection against direct electric shock, although RCDs should not be treated as a main and sole defence against direct contact injuries. This facility can be provided by fitting an MCCB on the MDA with sensitive earth leakage protection or by interposing a sensitive RCD in the supply from the MDA to the portable apparatus. The use of mains voltage portable apparatus, however, is not recognised by section 604 and should be discouraged.

To ascertain Zs, the designer needs to know Z_E, which is the earth loop impedance at the supply intake. The earth loop impedance between the intake and the relevant item of equipment can then be added, I_F calculated, and the suitability of the protective device ascertained from its time/current curves.

If the supply company provides an SNE (separate neutral and earth) service from its LV network, it should permit the use of its own earth terminal for the site earth and ought to be able to estimate the earth loop impedance at the intake to enable the designer to proceed with the calculations. If the service is CNE (combined neutral and earth) the supply company cannot legally permit the use of its CNE terminal for earthing and the contractor will have to provide a site earth electrode and terminal, in which case the value of Z_E cannot be determined without site testing.

A site earth will not provide as low a value of Z_E as the CNE terminal. In many cases the contractor will find it uneconomic to construct an earthing facility able to provide a low enough value of Z_E for the excess current protection to be adequate for earth fault protection. In these circumstances the remedy is to use an MCCB in the ISA fitted with earth leakage protection. As there is bound to be some earth leakage in the site installation, from capacitive currents, electric heating elements, through the insulation and from radio interference suppressors, the operational tripping current should not be set at too low a value, otherwise nuisance tripping of the whole installation may occur. Not less than 500 mA should be satisfactory. The MCCB should be capable of disconnecting the installation in not more than 5 s when the selected earth leakage tripping current is exceeded. For the majority of sites this arrangement will suffice, but for very large sites it may be considered prudent to sectionalise the system and provide local earth leakage protection capable of discriminating against the intake protection so as to shut down the relevant section, instead of the whole site, in the event of an earth fault. To attain this, faster operating times may be required for the section circuit breakers.

Where the supply company provides a site substation there should be no problem about using its earth terminal; Z_E will be very low and the excess current protection should be adequate for earth leakage protection except for the larger installations.

Confined conductive locations

In confined and conductive locations the potential electric shock hazard is increased and special precautions are needed. Examples of confined, conductive spaces are inside boilers and other metal vessels or inside metal pipes, flues and ducts where the area of body contact to earthed metalwork is likely to be substantial. Even if the interior is dry, the shock risk is enhanced, but if it is damp it is worse. In these circumstances the 110 V system is not considered safe and pneumatic, hydraulic or battery powered tools are advocated. For lighting, battery powered cap and hand lamps could be used or the luminaires could be supplied from a safety transformer at not more than 50 V. BS 7375 recommends that the 50 V transformer should be single-phase, double wound, with the mid-point of the secondary winding earthed so as to restrict the maximum shock voltage to earth to 25 V. Alternatively, the code of practice specifies a transformer with a 25 V unearthed output. The transformer should not, of course, be taken into the confined space.

Lighting

Site lighting may be divided between the fixed lighting provided for security and/or safe working and movement and local lighting used by individuals or small groups. In the first category there are floodlighting systems where the luminaires are mounted at high level on poles, cranes and other structures and usually operated at mains voltage. Robust, totally enclosed, weatherproof luminaires are preferable. Open types, where the lamp is exposed, should only be used where they are not vulnerable to damage from a carelessly handled scaffold pole or length of conduit, for example. During non-working hours, it is better to switch off the installation at the intake, but if it is necessary to leave any circuits energised they must be reasonably safe and proof against mischievous children, other vandals and thieves. Security lighting luminaires, for example, should be pole-mounted or fixed in relatively inaccessible positions.

Luminaires that are within reach and/or used and handled by site personnel should be operated on the 110 V system. Festoon lighting, unless out of reach, should also operate on the 110 V system. If operated at mains voltage, only the type with moulded-on lampholders, designed to seal against the glass bulb of the lamp, should be used. The detachable type which have spikes to penetrate the cable are not suitable because water may enter the joint and provide a conducting path between a live conductor and anyone handling the wet cable. Moreover, when the lampholder is removed, the holes left by the spikes again permit the ingress of water.

There are obvious difficulties in designing a lighting installation to cater for the changing needs on a construction site but, even so, good practice, as described in Chartered Institute of Building Services Engineers (CIBSE) Technical Publication *Building and Civil Engineering*, should be observed to ensure that the lighting contributes to, rather than detracts from, site safety.

Installation and maintenance

Safety is dependent on the quality of the installation work and the subsequent maintenance. If the initial installation work is competently carried out to the requirements of BS 7671 and a planned maintenance system inaugurated on its completion, the installation should be safe and will have a better chance of remaining safe for the duration of the project. On completion of the initial installation work the tests specified in BS 7671 should be carried out, including checks on the actual prospective fault levels (there is an instrument available for this purpose) to ensure that the circuit breakers originally selected, at the planning stage, are adequately rated.

A planned preventive maintenance system for the fixed installation should be drawn up and should include frequent visual inspections and combined inspections and tests (IEE Guidance Note No. 3 recommends a maximum of three months between the inspections and between combined inspections and tests). The aim is to detect damage and defects which may create danger and which should be repaired immediately. If repairs have to be deferred, the defective apparatus should not continue in use but should be disconnected from the system until it has been made good.

In addition to the fixed installation, all other electrical equipment and apparatus used on the site should be included in the preventive maintenance programme. This will included portable electrical tools, handlamps, lighting equipment and so on. By far the most important element of this maintenance is a routine visual examination of the equipment, which will detect most faults that can lead to danger. These examinations should be carried out as pre-use user checks and then periodically as part of a formal visual inspection, typically at a frequency of once every week for 230 V hand-held equipment and extension leads, and monthly for 110 V equipment and fixed 230/400 V equipment. Guidance on the frequency of these examinations, and the frequency of tests aimed at detecting defects, is published in HSE's Guidance Note HS(G)141 and the IEE's Code of practice for in-service inspection and testing of electrical equipment.

The persons carrying out the maintenance need to have the competencies to carry out the work safely and to detect failures or deterioration that may lead to injury. Whereas visual examinations do not need to be carried out by electrically-qualified persons, testing work carried out on the fixed installation should be carried out by electricians, or similarly qualified tradesmen.

Systems of work on the permanent installation

Work on the permanent installation is not electrically hazardous so long as it is not energised. Problems can arise, however, as a project nears completion and other trades ask the electrical subcontractor to energise the supplies to their apparatus although the electrical installation is not commissioned. Energising some circuits of an incomplete installation imposes additional responsibilities on the electrical subcontractor:

■ The responsible person must ensure that there are no accessible live parts, so live busbars in switchboards and distribution boards, for example, will need screening and will need to have warning notices displayed. Live terminals and cable tails in lighting points and socket outlet boxes could be a hazard to plasterers and decorators, so temporary covers will have to be provided together with notices warning them not to remove the covers or allow moisture to enter.

■ Only competent persons may work where there is danger, so young and/or inexperienced apprentices should only be employed in areas where there are no live conductors.

■ No work on or in the vicinity of bare live conductors should be done unaccompanied, so the contractor must ensure that nobody does such work out of sight and sound of other persons.

■ To ensure that circuits being worked on cannot inadvertently be made live they must be securely isolated and locked off.

■ Before handling any conductors on a partially energised system, operatives should use a potential indicator to prove the conductors to be dead.

It will be evident, therefore, that the electrical subcontractor should avoid providing temporary supplies from the incomplete permanent installation as far as possible, but should endeavour to satisfy such demands by providing supplies from the MDA(s). There could be difficulties about this if the temporary distribution equipment has been withdrawn or the contractual conditions require the apparatus to be connected to the permanent supply, to ensure correct phase rotation perhaps, before handover to the client. The problems can be minimised if the electrical subcontractor ensures that the power is readily segregated from the lighting installation and that the former is completed before the other trades need it.

Existing installations

It is more difficult to avoid electrical safety problems on alteration, extension and refurbishment work if the existing installation has to be kept energised. Adequate safety rules and work planning are not only legally required to comply with the CDM Regulations but also are essential if the hazards of live working are to be contained. The effort is usually economically worthwhile, apart from the safety gain.

SAFETY RESPONSIBILITY

The responsibility for site safety is complicated because of the number of different firms present, each of which will have its own safety policy and some of which will have their own safety rules. The CDM Regulations allocate safety responsibility, with the 'principal contractor' and the 'planning supervisor' having key roles in coordinating health and safety during the planning and construction stages of a project. In particular, they should lay down the site safety rules and enforce them.

The principal contractor normally provides the temporary electrical installation. If he has an 'in house' electrical department, he may well install and maintain the installation and, in such a case, he is responsible for its safety. The electrical staff member in charge of the site installation will exercise the technical responsibility. This official should ensure that the planned maintenance is done and that defects and breakdowns are reported and rectified immediately. Some main contractors hire the distribution equipment and arrange for the hirer to install and maintain it. As the occupier, they still retain responsibility for the safe use of the electrical installation, but responsibility for its maintenance in a safe condition rests on the hirer who should appoint competent, electrically qualified persons to do the work.

Responsibility for the safety of connected apparatus generally rests on whoever has responsibility for its presence on site. The principal contractor, however, would be well advised to stipulate, in the site safety rules and perhaps in the subcontracts, that any electrical apparatus or electrically powered equipment brought on site must be serviceable and safe. Arrangements must be made for its safe installation and maintenance and for users to be trained to use it safely. The rules should also prohibit the use of apparatus that becomes defective, until the defect has been repaired.

If the electrical subcontractor energises part of the permanent installation before commissioning and handover to the client, he is responsible for its safety and must take the precautions already described. Although he is not responsible for the safety of the connected apparatus installed by other trades, he has a duty to notify them both before energising and before switching off the supply to their apparatus.

Chapter 12

Underground Cables

INTRODUCTION

Every year there are thousands of incidents involving damage to buried cables, some of which result in burn injuries and a few (typically one or two) are fatal. Most of the incidents occur during street works where utility companies, construction companies and local authorities carry out excavations in the highway and damage an electricity supply company's underground mains. Other incidents take place on construction sites and other private property and involve a supply company's service cables or mains, or distribution cables belonging to the site occupier.

The burn accidents invariably result from manual excavation when, for example, an operator using a pneumatic drill penetrates a concealed, buried cable with the tool. A short circuit will occur when the bit comes into contact with one or more of the live conductors and the metallic armouring and/or sheath of SNE cables or the CNE conductor where a PME distribution system is in use. The initial phase/earth or phase/neutral fault usually develops and involves the other phases, causing arcing which may emerge as a flame arc and blast from the hole made by the tool, injuring the operator.

The electricity supply companies' low voltage mains are frequently involved because there are more of them and because they are not well-protected against excess current faults. The protection tends to be slow in operation, or fails to operate at all, and the arcing persists until the fault burns itself clear. The number of burn injuries resulting from HV distribution cables being struck tends to be lower than the injuries from LV cables being struck. This is because there are fewer HV cables, because HV cables are usually buried at a greater depth and are more likely to be physically guarded, with cable tiles for example, and because the protection on HV cables operates rapidly so that the fault does not develop and little arcing occurs.

There are few reported accidents where the victim received an electric shock. This is probably because the fault involves an earthed conductor so the tool does not attain a significant voltage to the surrounding ground and because very often the operator is insulated from the ground by footwear such as rubber-soled boots.

In the past, there were fewer accidents because buried cables were better protected. Most of the supply authorities used paper-insulated, lead-sheathed, single-wire or steel-tape armoured and served cables and very often protected them with cable tiles. For economic reasons cable tiles are seldom used now and many of the mains laid in recent years are unarmoured CNE cables with plastic insulation and sheathing, such as PVC and cross-linked polyethylene (XLPE). The outer concentric conductor, which is the earthed neutral, is copper or aluminium tape and consequently more vulnerable to damage than a steel-wire-armoured cable. Every year more new mains cables of this type are laid so the resultant hazard to excavators is growing and more accidents can be expected to occur unless the construction industry becomes more effective at taking precautions against striking buried cables.

A considerable effort has been made to contain these accidents. There has been no lack of safety advice from the HSE and the supply companies. In 1977 the public utilities formed a committee called the National Joint Utilities Group (NJUG) which studied the problem and issued a series of safety guidance booklets, and some of the larger civil engineering contractors established safety training schools for their staff, but the accident rate is still far too high. The HSE's main guidance document on the topic is HS(G)47 *Avoiding danger from underground services*, which provides comprehensive guidance on the systems of work that should be adopted to prevent underground services being damaged.

SAFETY PRECAUTIONS

The precautionary safety principles are fairly simple and comprise:

(1) Only trained personnel should organise and carry out excavation work in the highway or anywhere else where there could be buried cables.

(2) No work should start until location drawings of the buried cables have been obtained.

(3) The position of the buried cables shown on the location drawings should be checked with a cable location tester.

(4) The route of the cables should be marked on the surface of the ground.

(5) Further markings should be provided 0.5 m on either side of the cable route to indicate the danger zone.

(6) Any excavation necessary within the danger zone must be carried out with great care to avoid damaging the cable.

(7) The exposed cable must be supported where necessary and protected from damage.

(8)   Reinstatement of the cable should be in accordance with the instructions of the owner.

Training

The training referred to in (1) above can be in-house or external but should be undertaken by all personnel before they undertake any excavation work where there may be buried cables. The training should embrace the legal requirements such as Regulation 14(c) of the Electricity at Work Regulations 1989, the risk assessment requirements of the Management of Health and Safety at Work Regulations 1999, the CDM Regulations, the Construction (Health, Safety and Welfare) Regulations 1996 and the Health and Safety (Consultation with Employees) Regulations 1996.

The training should also cover how the hazard arises; the type of burn injuries that can occur; accident statistics; accident and dangerous occurrence reporting requirements; cable laying practices including types of cable, depth of laying, physical protection and locations; cable location plans and how to obtain them; permit to work practices and procedures; and the use of cable locators. The practical work should demonstrate the use of the cable location plans and cable locators, marking the cable route and the danger zone, the use of power tools in and near the danger zone, digging trial holes to locate cables, the use of hand tools to expose the cable, protection of the exposed cable, and backfilling precautions.

Drawings

As regards point (2) above, it is not sufficient to contact just the local electricity company's district or regional office to obtain cable plans as there may be buried cables belonging to other organisations, such as cables supplying street lighting and other 'street furniture'. It is always advisable, therefore, to make enquiries of power transmission companies, the local authority for street lighting cables and, in the case of private property, the site owner and/or occupier. In some regions there are coordinated single-point-of-contact services covering information on buried utilities.

It should be borne in mind that the plan may not be entirely correct as it can only reflect the knowledge available to the cable owner and in many cases will show the cable location as originally laid. Since then excavation work may have resulted in an unauthorised displacement of the cable by others who may not have notified the cable owner, or the ground may have been regraded thus altering the cable depth, or reference points showing the cable location dimensions, such as kerb lines, may have been changed. References on the drawing to cable tiles or marker tapes should be treated with caution, as subsequent excavation work may have displaced such protective devices. If the type of cable is not specified on the drawing the owner should be asked for the details.

Under the New Roads and Street Works Act 1991 the local authority has to keep a register of information concerning the apparatus in the streets and coordinate the activities of excavators digging them up. The excavator should therefore be able to obtain all the details of the buried services from the local authority without the need to approach all the undertakers before excavating.

Cable locators

There is a variety of cable locators readily available, varying in sophistication but of three generic types:

(1) those that detect the magnetic field of a live cable;
(2) those that detect the very low radio frequencies reradiated by long-buried metallic structures; and
(3) those that detect metal.

The third type is of limited value in urban area highways because there is a great deal of buried metal of other services present and so it is almost impossible to identify the cable sought. Another type of detector, the ground mapping radar, is becoming increasingly available and can be used to map out the distribution of underground services.

The types that detect the magnetic field are most effective when a cable carries a substantial current because the field is proportional to the current. Cables that are live but carrying little or no current generate insufficient magnetic field to be readily detectable – this fact is frequently ignored or not understood and cables that are live but off-load are often not detected until they are struck and somebody injured. This problem can be overcome by injecting a signal into the buried cable and signal generators are available for this purpose; this technique, of course, relies upon knowing that the cable is there in the first place, which is not always the case.

Cable locators should be used only by personnel trained in their use. It is good practice to use them from time to time as the excavation work proceeds until the cable is exposed. Quite often, and particularly in urban areas, there may be more than one cable laid at the same or different depths. Care in the use of the cable plans and locators is needed to ensure that the location of all the cables is determined and that cables located in close proximity to other cables, including above other cables, are properly resolved. Cables are not always distinguishable by sight from other services, but the use of a locator should enable the operator to be sure.

Cable route marking

On paved surfaces the cable route and danger zone lines can be chalked or painted, but elsewhere it will usually be necessary to delineate with pegs driven into the ground. Wooden pegs are safer than metal as they are less likely to damage the cable.

A number of serious accidents have occurred when the cable markings have been covered up. For example, a construction company was laying a road and pavements in a new housing estate. The principal contractor had located an 11 kV buried cable and marked out its route and depth using yellow paint sprayed on to the ground surface. After a few days the ground was covered by a subcontractor with a shallow layer of hard-core as a base for the road and pavement – the hard-core was laid over the yellow markings. A labourer then came along to drive short lengths of steel reinforcing bar into the ground that would act as supports for the formers for the concrete on to which the kerb stones would be placed. He had no knowledge of the buried cables and, of course, never saw the yellow markings which were underneath the hard-core. He drove one of the lengths of bar into the buried cable; the bar was blown back from the fault and hit him in the face causing injuries that resulted in the loss of his left eye, and he suffered burn injuries to his chest and face. The moral of the story is, perhaps, obvious.

Excavation

Where a substantial amount of excavation is required in the danger zone, it is advisable to dig one or more trial holes to find the cable and then work along it from the trial holes to expose it. Careful manual methods should be used to avoid damage to the cable. Surface construction may require the use of power tools to break up the surface for removal, in which case it is advisable to fit the bit with a collar to prevent deep penetration and no drilling should be done immediately above the cable. Sharp-pointed tools such as picks should not be used immediately adjacent to the cable even if it is armoured, as the pick points are capable of penetrating between the strands of wire armour and through badly corroded steel tape.

The sharp edges of shovels can damage lead and aluminium sheaths and the outer copper or aluminium conductors of CNE cables. If a cable is damaged, work should cease in the immediate vicinity of the damage, a temporary barrier, with a warning notice, should be erected to keep people away, and the cable owner asked to repair the cable before work is resumed at the damage site. It is imperative that nobody inspects the cable until the owner confirms that it has been made safe.

As far as the person carrying out the excavation is concerned, it is safer to do the excavation using a mechanical digger with a back-hoe or similar implement. If the cable is struck, the person in the vehicle's cab is unlikely to be injured. Any banksman or other labourer in the vicinity of the digging tool may suffer an injury, though, from the explosion products of cable strikes, and this does happen occasionally. Not unnaturally, cable owners condemn the use of excavating machinery in the vicinity of their cables because of the likelihood of damage but, on the other hand, the use of such machinery does save accidents. It does, however, increase the number of dangerous occurrences.

Cable safeguarding and reinstatement

Exposed cable spans of more than about 1 m should not be left unsupported and such spans should not be used as steps or to support anything. It is better to avoid moving cables because of the danger of damaging them inadvertently. Old cables may have embrittled sheaths and/or corroded armour and are vulnerable to damage if moved. Cables that have been displaced should be secured where necessary to prevent them from falling back into the excavation, and suitable guards should be provided to prevent damage to them. The cable owner's advice on reinstatement should be sought and followed. If this is unavailable the contractor should level the bottom of the trench and firm it up to prevent subsequent subsidence.

The cables should be relaid in a bed of sand or riddled earth, free from sharp stones and at their original spacings from each other and from other services. The cables should be handled with care, avoiding unnecessary bending and strain. They should be covered with sand or riddled earth and any tiles or warning tape replaced above them. Where there is sufficient slack, the cables should be snaked uniformly to alleviate possible strains in the event of subsidence. If the cables have been displaced from their former positions, when they are relaid the cable owner should be notified before backfilling so that the location plans can be amended.

Chapter 13

Safety-related Electrotechnical Control Systems

INTRODUCTION

This chapter is concerned with electrotechnical control systems that perform safety functions; such systems are commonly known as safety-related control systems. In this context, the term 'safety function' refers to a control action that needs to be taken to ensure that the risk associated with a particular hazard is reduced to a level that is acceptable or tolerable. For example, an interlocking circuit associated with an interlocked guard protecting access to live conductors inside a test rig may work to ensure that the conductors are made dead when the guard is opened and that the conductors remain dead while the guard remains open. This interlocking facility performs a safety function. Whilst the subject does not strictly come within the scope of 'electrical safety', the chapter is included because many electrical engineers, technicians and electricians are required or expected to have an understanding of control system technology. The purpose of the chapter, therefore, is to provide an insight into the main devices used and the principles that underpin the achievement of adequate levels of safety.

So as to limit the scope of the chapter to a manageable level, only applications associated with machinery safeguarding and the safeguarding of electrical hazards are addressed in detail. This means that highly specialised areas such as safety-related systems for emergency shutdown and instrumented protective functions in petrochemical and chemical plants are not considered, although the principles will be the same as those for the machinery sector. The fact that the chapter addresses electrotechnical control systems means that the design of systems that are not based on electrical, electronic and programmable electronic technologies is not covered.

A generic electrotechnical control system used in machinery and other applications is depicted in Fig. 13.1. This shows that at the heart of the system is the 'processing unit' (often called a 'logic solver'), the subsystem that processes data and signals from input devices and sends signals to output devices. The processing unit may be implemented in a variety of

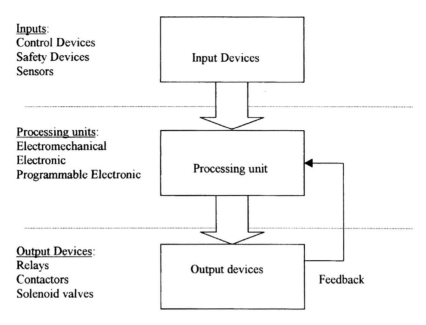

Fig. 13.1 Generic block diagram of a control system.

technologies, ranging from electromechanical devices such as relays and discrete timers, to electronic systems, to programmable electronic systems such as programmable logic controllers (PLCs). The inputs to the processing unit will typically emanate from:

■ Control buttons and switches for operator-initiated functions such as 'start', 'stop', 'mode change', 'inch' and 'pause'.
■ Sensors on a machine which detect the position and rate of change of position (speed) of parts of the machine (including guards) and the product being handled by the machine. These will be limit switches, photoelectric sensors, capacitance switches, magnetically-operated switches, and so on.
■ Sensors that detect parameters such as temperature, pressure, level and flow.
■ Other control systems on linked machinery.

The outputs from the processing unit will typically interface with electro-mechanical contactors, relays, variable speed drives, solenoid-operated pneumatic valves, and indicating light bulbs.

The chapter describes the essential elements of electrotechnical safety-related control systems and identifies the principle CEN and CENELEC 'A' and 'B' standards that cover them. It also describes two standards that are having an important influence in the field: BS EN 954 Part 1 Safety of machinery – Safety-related parts of control systems, and IEC 61508 Functional safety of electrical/electronic/programmable electronic safety-related systems. In association with this, advances in safety technology, such as the development of 'safety programmable logic controllers' and the use of fieldbuses to carry safety signals, will be described.

ACCIDENTS AND INCIDENTS

A minority of accidents in the machinery sector are the result of control systems failures. Most are the result of machinery operators and maintenance staff entering hazardous areas by removing or bypassing guards and other safeguarding devices without having isolated power from the actuators that cause dangerous movement. For example, operators clearing blockages on machinery frequently intervene in the machine with the power still applied, clear the blockage and then get hurt when the machine moves. And maintenance staff carrying out fault finding work with the power still applied often cause sensors to change state while they are working in the hazardous zone and then get hurt when the machine responds by moving unexpectedly.

However, whereas unsafe systems of work dominate the machinery accident statistics, it is important that the safeguarding systems and the safety-related control systems associated with them, are properly specified, designed, built and maintained so as to maximise safety.

BASIC PRINCIPLES

The electrotechnical safety-related systems being considered generally aim to safeguard against exposure to electrical hazards (shock and burn injuries and the effects of explosions and fires that have an electrical origin); mechanical hazards (crushing, shearing, cutting, puncturing and similar injuries); thermal hazards associated with parts and liquids or gases that are at a high temperature; and other hazards that may occur during the operation, setting and maintenance of the equipment under control. Many of the systems are referred to as interlocking systems because they provide an interlocking function, usually between a guard and a hazard, such that when a guard is opened the hazard protected by the guard is removed or, alternatively, the hazard must be removed before the guard can be opened. However, interlocking functions are not, by any means, the only safety functions that need to be addressed. Overspeed control on paper and steel mill machinery, limitation of temperature in thermoforming machines, emergency stop systems, and anticollision systems on roller coasters are all examples of safety functions implemented in electrotechnical technologies that are outside the 'interlocking' category.

The legislation and standards covering machinery safety stipulate a hierarchy of measures that should be taken to protect against the hazards identified above. In the first place, efforts must be made to design them out. If this cannot be done, they should be safeguarded using fixed guards and barriers. Next in the hierarchy is the use of interlocking and trip devices, which make use of the type of safety-related control systems being considered here. Finally, and least acceptable, is the provision of instructions and information to people who may be exposed to those hazards that remain.

As a general principle, safety-related control systems such as interlocks should be well designed, be of simple construction, and be sufficiently robust for the application. Furthermore, it should not be possible easily to defeat interlocking devices, yet where appropriate it should be possible for authorised persons to override them for tests or other necessary purposes.

A particularly important requirement for such systems is that they must have a safety integrity matched to the amount of risk reduction that the system is aiming to achieve, where the term 'safety integrity' is usually understood to be a 'goodness' factor that combines the concepts of reliability and fault tolerance. This may sound a difficult concept at first, and its realisation is often far from easy, but the principle is straightforward. It simply says that the higher the contribution that a safety-related system makes to safety, the better must be the system's safety performance. Safety performance of such systems is considered later in the chapter, after the main elements shown in Figure 13.1 are described in more detail.

LEGAL REQUIREMENTS

The general provisions of the Health and Safety at Work etc. Act apply to the safety of control systems, both in terms of their supply and their use. Whereas the Act is goal-setting and non-prescriptive, there are Regulations that provide more specific legal requirements.

In the context of safety-related control systems, the Essential Health and Safety Requirements of the Supply of Machinery (Safety) Regulations 1992 lay down generic requirements that must be considered by suppliers for the safety and reliability of control systems; control devices; starting and stopping devices; mode selection; failure of the power supply and the control circuit; software; and movable guards.

Users of machinery and other work equipment need to be familiar with the

Provision and Use of Work Equipment Regulations 1998 and the associated Approved Code of Practice. These Regulations have a number of provisions relating to control systems that tend to mirror those in the Essential Health and Safety Requirements of the Supply of Machinery (Safety) Regulations. Indeed, the Regulations impose a duty to provide and maintain suitable work equipment that conforms to standards specified in other legislation implementing European Directives, including the Machinery Directive. The relevant Regulations are:

- Regulation 11 Dangerous parts of machinery
- Regulation 14 Controls for starting or making a significant change in operating conditions
- Regulation 15 Stop controls
- Regulation 16 Emergency stop controls
- Regulation 17 Controls
- Regulation 18 Control Systems

MAIN STANDARDS

The overarching standard for machinery safeguarding is BS EN 292 Parts 1 and 2 Safety of machinery – Basic concepts, general principles for design. This standard sets out the general principles and is broadly equivalent to the excellent machinery safeguarding standard, BS 5304, produced by BSI in 1984. BS 5304 remains a first class reference document, if only because the illustrations in it, and the descriptive nature of its text, make it far more user-friendly than any of the newer European harmonised standards. Against that, the standard is beginning to show its age because it does not reflect recent advances in technology.

The other relevant main A standard is BS EN 1050 Safety of machinery – Risk assessment, which provides advice on how to carry out a risk assessment on machinery.

In the context of this book, the following B standards address the main electrical and control systems issues on, and systems used in, machinery safeguarding (in addition to the previously mentioned BS EN 954-1 and the IEC standard IEC 61508):

- BS EN 60204-1　Electrical parts of machinery
- BS EN 1088　Interlocking devices associated with guards
- BS EN 61496　Electrosensitive protective equipment
- BS EN 418　Emergency stop devices
- BS EN 574　Two hand control devices

The techniques described in these standards are aimed at machinery safe-guarding but the principles also apply to control systems, particularly interlocking systems, which are used to safeguard electrical hazards.

INPUT, INTERLOCKING, TRIP AND EMERGENCY STOP DEVICES

There are many types of devices used in safety-related systems that provide inputs to the processing unit, with the main ones described in the following paragraphs. Some of these act as interlocking devices, usually associated with guards. These devices – limit switches, tongue-actuated switches, guard locking devices, proximity and magnetic switches – are described in BS EN 1088. Other devices act as presence sensing and trip devices – photoelectric light curtains being an obvious example – and are generally covered by B standards that are specific to the type of device, an example being BS EN 61496.

Other devices measure parameters such as speed, acceleration, pressure, level, flow and temperature. These types of devices are not described in this book but the main technologies are:

- *Flow measurement:* flow meters using ultrasonics, electromagnetics, turbines, and pressure differential measurement.
- *Temperature measurement:* thermometers, thermistors, thermocouples, radiation pyrometers and bimetallic strips.
- *Force measurement:* spring balances, strain gauges, piezoelectric transducers.
- *Pressure measurement:* force-balance and elastic element systems, and solid state pressure transducers.
- *Velocity measurement:* tachogenerators.
- *Acceleration measurement:* seismic-mass, strain-gauge, potentiometric, piezoelectric and servo accelerometers.

Limit switches

Limit switches are in widespread use in interlocking systems and are also used to indicate the position of moving parts. A typical limit switch used in machinery safeguarding applications is depicted in Fig. 13.2, which shows a switch with a roller plunger actuator. A switch of this type is portrayed in Fig. 13.3, which also shows a switch with a straightforward plunger actuator. The figure also shows an example of a typical contact arrangement in these switches. In this case, the switch has two normally open (NO) and two

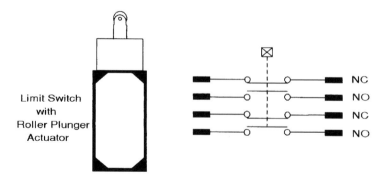

Fig. 13.2 Typical limit switch and contact configuration.

normally closed (NC) contacts, although manufacturers offer many other combinations of contacts.

The contacts in limit switches can be wired into different parts of control circuits, but it is usual to wire normally open contacts into primary safety circuits. This is because safety functions are frequently best implemented on deenergisation to ensure, for example, that open circuit faults do not lead to dangerous failures and that the system moves to a safe state in the event of a power failure.

In the context of safety applications, there are two generic types of limit switch, known as positive mode and negative mode switches. These are shown in Figure 13.4, which illustrates the switch's plunger being moved by a rotating cam. The positive mode switch is safer than the negative mode

Fig. 13.3 Two limit switches, one with roller plunger actuator and one with plunger actuator.

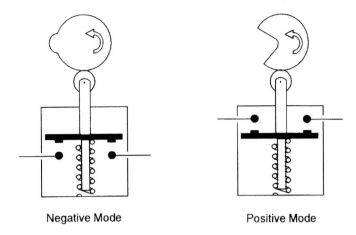

Negative Mode   Positive Mode

Fig. 13.4 Positive and negative mode operation.

switch because its contacts are positively forced apart by the rotating cam, whereas the negative mode switch relies on spring pressure to open the contacts. The negative mode switch could fail to danger if the spring were to break, or if the contacts were to weld together, or if the plunger were to become stuck. Moreover, it can easily be defeated by taping or wedging the plunger down.

Limit switches, and other interlocking devices, can either be inserted in the power supply circuit (called power interlocking) or in the control circuit (called control interlocking), depending on the size of the current to be interrupted and the rating of the switch. In general terms, power interlocking is safer than control interlocking because the former acts directly on the supply whereas the latter relies for successful operation on interposing components and therefore has the potential to be less reliable; for this reason, power interlocking is frequently used in high risk applications such as plastic thermoforming machines and power presses where opening a gate or guard leads directly to disconnection of the power source. However, control interlocking is more commonly used than power interlocking because in many applications it is undesirable simply to switch off the supply to the machine and because of the current breaking limitations of the interlocking switch.

In general terms, interlocks must not be easily defeatable. Figure 13.5 shows a limit switch used in the interlocked cover of a test jig in which a component to be tested is first inserted in the jig and becomes automatically connected to the power supply via the test contacts. Closing the lid allows the limit switch to complete the control circuit, which energises the coil of the contactor leading to the contactor's contacts closing and the test contacts

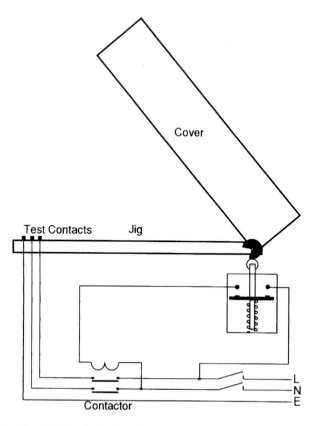

Fig. 13.5 Test jig with interlocked enclosure.

becoming live. Opening the lid causes the limit switch contacts to be forced apart by positive action, thereby deenergising the contactor coil, causing the contacts to open and the test contacts to become not live. This means that the test contacts will only be live when the lid is closed. Moreover, the limit switch is not easily defeated so as to switch the test contacts on while the lid is open.

A possible dangerous failure mode in this type of design is the failure of the contactor with its contacts in the closed position, meaning that the test contacts would remain live with the lid open. A possible failure mechanism would be the contacts welding together if excessive current is passed through them. This can be prevented by ensuring that the contactor is of good quality, is built to a recognised standard, is properly rated for the duty, and by inserting a fuse or a circuit breaker in the supply to provide adequate over-current protection.

Although the discussion so far has concentrated on safeguarding access to live parts, limit switches also have important applications in machinery safeguarding. The cam-operated switch configuration shown in Fig. 13.5, for example, is commonly used on hinged guards that prevent access to mechanical hazards on machinery. Roller plunger limit switches are often used in association with sliding guards, as depicted in Fig. 13.6. Movement of the guard causes the shaped ramp to push open the positive mode contacts of the switch, opening the supply to a contactor and thereby switching off the prime mover such as an electric motor.

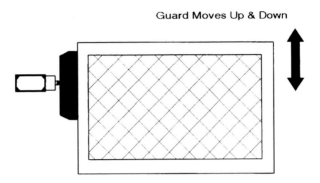

Fig. 13.6 Limit switch operated by ramp on sliding guard.

Positive and negative mode switches are often used together in safe-guarding applications to prevent common mode failure caused by, for example, the displacement of a guard. An example of this configuration is shown in Fig. 13.7, which depicts a sliding guard with a ramp attached; the ramp operates the plungers of the two switches. If the guard were to become displaced, it is most likely that one of the two switches would operate. Provided that the system is properly maintained, a single fault is unlikely to cause a failure to danger of the interlocking system.

Another example of the application of limit switches is in lift safety cir-cuits. These employ hardwired components connected together in series in such a way that they must all be closed before the lift can operate, as shown in Fig. 13.8. If any one of the components were to become open circuit, the safety circuit would cause the main up or down contactors to open and the lift to stop. The ultimate top and bottom switches in the lift shaft are limit switches, which have contacts that open if the lift overshoots by a predetermined distance the top or bottom floors.

Tongue operated switches

A special type of limit switch in common use in machinery safety applications is the tongue-operated switch, an example of which is shown in Fig. 13.9. In

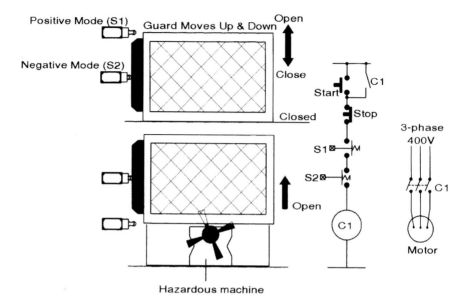

Fig. 13.7 Positive and negative mode limit switches used in guard interlocking.

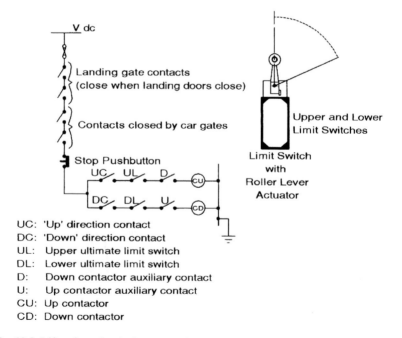

UC: 'Up' direction contact
DC: 'Down' direction contact
UL: Upper ultimate limit switch
DL: Lower ultimate limit switch
D: Down contactor auxiliary contact
U: Up contactor auxiliary contact
CU: Up contactor
CD: Down contactor

Fig. 13.8 Lift safety circuit, incorporating roller lever switches for upper and lower ultimate limit switches.

Fig. 13.9 Tongue-operated limit switch.

this case, a shaped metal tongue that enters or leaves an aperture in the body of the switch positively operates the internal contacts both when being inserted into and withdrawn from the switch, thereby minimising the chance that the switch contacts could stick together. The tongue would be attached, for example, to a moveable guard and the body of the switch would be attached to the frame of the machine, as depicted in Fig. 13.10. It is possible to defeat this type of switch by inserting a spare actuating tongue, so any spare tongues should be kept locked away or thrown out to prevent this happening. An alternative means of defeat is to remove the tongue from the machine and to leave it inserted in the switch. To prevent this happening, the tongue should be securely attached to the machine and should only be capable of being removed by using a special tool – in this context, a 'special tool' does not include common screwdrivers, Allen keys, wrenches and so on.

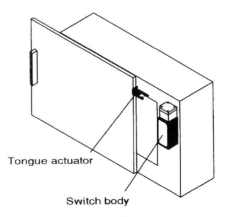

Tongue actuator

Switch body

Fig. 13.10 Application of tongue-operated limit switch.

Switches with guard locking

An advantage of the tongue-operated switch is that the switch mechanism can be designed to lock the tongue in place. This is commonly achieved using a spring-applied catch that is released by a solenoid-operated bolt. An example of this type of device, manufactured by EJA Ltd, is shown in Fig. 13.11, which shows the device with the tongue actuator both in and out of the body of the switch. The particular model shown has a key-operated locking facility.

Fig. 13.11 Example of tongue-operated interlocking switch with guard locking, with actuator both out of and inserted in the body of the switch.

The tongue actuator is inserted in the switch as normal, when the guard to which it is attached is closed, positively operating a set of normally open and normally closed contacts that are wired into the machinery control circuit. The tongue is automatically locked in place by a mechanical latch so that the guard cannot be opened. The tongue can only be unlatched by energising a solenoid inside the switch – this controls a plunger which acts on the latching mechanism. The solenoid will normally be energised by the machinery control system when a safe state has been achieved, such as when hazardous rotating parts have stopped or when power has been removed from normally-live parts.

In the case of the device shown in Fig. 13.3, when somebody enters the hazardous area the key switch can be turned to the 'lock' position and the key removed. This prevents the switch from being operated again until the key is replaced and turned to the 'unlock' position, thereby providing the person entering the hazardous area with control over the interlocking system.

Time delay switches

One of the main purposes of the interlocking switch with guard locking is to prevent a guard being opened until the hazard safeguarded by it has been removed. Another technique that achieves the same end in the case of rotating machinery is a time delay switch, an example of which is illustrated in Fig. 13.12, which shows the switch attached to a centrifuge. The time delay switch has a long threaded bolt with a hand-operated knob at the end. To release the guard, the operator has to unscrew the bolt. During the first few turns, the plunger of the limit switch is forced down as it moves out of the indentation in the bolt and the switch contacts are opened, causing the control circuit to become deenergised. The bolt must then continue to be unscrewed to allow the hinged lid to be released and this takes sufficient time to ensure that the machinery has come to a halt and the hazard has been removed before the lid can be lifted. The internal parts of a typical time delay switch are shown in Fig. 13.13.

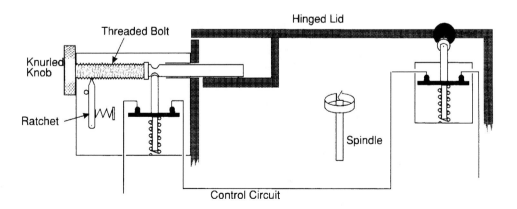

Fig. 13.12 Centrifuge with two interlocking switches closed in the run mode.

Fig. 13.13 Internal parts of typical time delay switch.

Non-contact switches

There is a variety of techniques that can be used to produce switches that have a non-contact mode of operation. By far the most common are devices that employ magnetic coupling.

Magnetic switches have two components: a unit containing a magnet and a unit containing a switch that can be operated by the magnet. In safeguarding applications, it is normal to mount the magnet unit on a movable guard and the switch unit on the fixed part of the machine, as depicted in Fig. 13.14. For correct operation, the magnet must be correctly aligned with the switch unit. Bringing the two close together, typically to within 5–10 mm, will cause the switch to close. If it is connected into a control circuit, this can then be used to enable machinery movement. Only magnetic switches that are specifically designed for safety applications should be used – they are designed so as to be resistant to vibration and with overcurrent protection to prevent the switch contacts welding together.

Bringing a magnet up to the switch unit could defeat it, so it is important that the switch unit is mounted so that it is not easily accessible when the guard is open. In most modern designs, the magnetic coupling is designed to reduce the likelihood of the switch being defeated in this way. In one design,

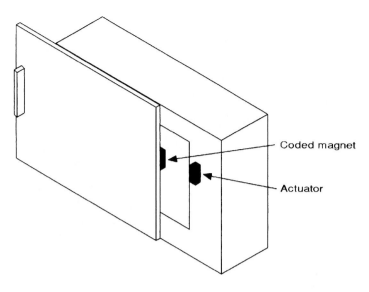

Fig. 13.14 Magnetically operated safety switches on sliding guard.

the magnet has to be inserted into the body of the switch for it to operate. The hole in the switch is made a complex shape so that only a magnet of matching shape can enter, meaning that a simple bar magnet cannot be used to defeat the interlock. In another design, the system has two channels and features an anti-tamper code that ensures that it can only be operated by its own magnetic actuator unit.

In another design of non-contact safety switch, a coded signal is trans-mitted between the two heads (one on the fixed part of the machine and one attached to the guard) when they are in close proximity. The code is unique to the particular unit, making it very difficult to defeat.

One of the advantages of non-contact safety switches is that the assemblies are hermetically sealed against the ingress of moisture and particles. This makes them especially useful for applications where hygiene may be an important issue, such as in food processing machines where food particles may be able to enter and decay inside the actuator and plunger openings of other types of forced contact switches. Their main disadvantage is that the switch contacts are not physically forced apart, so they are not suitable for use on their own in applications that demand a high level of safety integrity, as will be discussed further in a later section of this chapter. However, the level of integrity can be improved by using redundancy and monitoring techniques, and devices suitable for this are supplied by many of the safety component manufacturers.

Captive-key and trapped-key devices

A captive key system comprises a switch and integral lock, typically fitted to a machine or equipment enclosure. The switch is operated by inserting a key into the lock, with the key being secured to the moving guard or enclosure door, usually as an integral part of a handle. When the guard or enclosure door is closed, the key enters the lock. Turning the handle then turns the key that operates the switch. The switch can therefore only be operated by closing the guard or door and turning the handle, locking the guard or door in its closed position.

A trapped key system has a single key that is transferred between a lock on a guard or enclosure door and a key-operated switch incorporating a lock on a control panel. The key is trapped in either the lock on the guard or the lock on the control panel. The key can only be released from the guard when the guard has been closed and locked. Once it is released from the locked guard, the key can be inserted in the control panel's lock and then turned to switch on the control function. At this point, the key becomes trapped in the lock. The key must then be turned before it can be released to open the guard. This has the effect of turning off the control system, removing the hazard covered by the guard. There are many variations on this theme, some of which involve a multiplicity of keys and locks to ensure that all guards are closed and/or all power sources removed before guards can be opened.

An example of an electrical system that uses more than one key is shown in Fig. 13.15. The system is used to prevent the paralleling of transformers feeding a low voltage switchboard, which is divided by bus-section switches. In normal operation the bus-section switches are open and each switchboard section is fed from its own transformer, so the feeder switches are closed. All switches are provided with locks and there are three keys that are free when the switches are locked open but trapped when the relevant switches are closed. Key X will only fit its feeder switch and the adjoining bus-section switch XY and key Z will similarly fit only its feeder switch and the adjoining bus-section switch YZ, but key Y will fit both its own feeder switch and either of the bus-section switches XY and YZ. This arrangement permits the following operational choices:

(1) *Three feeder operation.* Bus-section switches open and switch-board fed from its own transformer.
(2) *Two feeder operation.* Bus-section XY closed, bus-section YZ open and sections X and Y fed from either X or Y. Section Z is fed from feeder Z.
(3) *Two feeder operation.* Bus-section XY open, bus-section YZ closed and sections Y and Z fed from either Y or Z. Section X is fed from feeder X.
(4) *One feeder operation.* Both bus-sections closed and whole switchboard fed from only one of the three feeders.

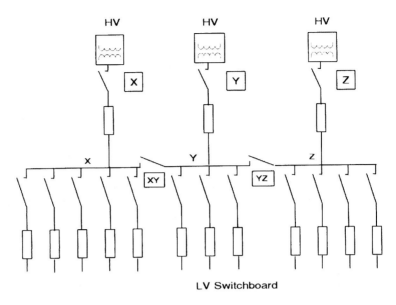

Fig. 13.15 Trapped key interlock system on a feeder switchboard to prevent paralleling of the transformers.

Two-hand control devices

A two-hand control device requires activation of two control buttons to initiate a control action. Most devices are synchronous two-hand control, meaning that the design of the system must ensure that the buttons need to be pushed within 0.5 s of each other, so that they can only be actuated by a single person. Operation of the two buttons by a single hand is prevented by, for example, spacing the buttons at least 550 mm apart and/or by shielding them. The principle is that these features will result in the operator's hand being kept away from the danger zone. It does not, of course, prevent another operator accessing the danger zones, which means that the machine must be built and laid out so as to prevent this happening.

Trip devices

Some safeguarding devices are used to detect the presence of people and to trip the machines they safeguard. The main two to be described here are photoelectric safety systems and laser scanners, which are both types of active optoelectronic electrosensitive protective systems. Pressure sensitive devices such as pressure mats and trip bars are other types of trip device.

Photoelectric safety systems

The generic name used in standards for photoelectric safety systems, and similar devices, is electrosensitive protective equipment (ESPE). These systems generally comprise a transmitter unit and a receiver unit between which a light curtain or a set of beams is established. Breaking the light curtain or beams causes the control unit to open the switch or switches, known as output signal switching devices, on its output. These switches are connected into the control system of whatever machine the light curtain is protecting, so that opening the switches causes the machine to stop or some other action to be taken.

The detection capability of light curtains must be specified by the manufacturer so as to describe the minimum size of object that will cause actuation of the system. In many applications, the light curtains must be actuated by objects with a size in the same order as that of a human hand, although in some applications larger object detection capability may be appropriate. Light curtains have many safety applications, but they are commonly used to safeguard dangerous parts where frequent operator access to hazardous areas is necessary, such as the blades of paper-cutting guillotines and the tools of power presses. In these types of application, the detection capability of the device must be objects the equivalent size of a human hand or even smaller. They are also used to protect apertures in perimeter fencing where products have to be fed in or taken out, examples being palletisers and depalletisers, manufacturing robots and packaging machines. In these latter applications, the detection capability may be enlarged because it is whole-body or part-body access that is being prevented. The two types of application are depicted in Fig. 13.16.

An example of photoelectric devices being used to safeguard electrical hazards can be found in high voltage meat stimulators. These devices, depicted in Fig. 13.17, are increasingly being used to tenderise meat by applying a pulsed voltage of about 800 V to the carcasses of animals such as pigs, cattle and sheep shortly after slaughter. Essentially, the carcasses are suspended from a chain conveyor that forms one pole of the electrical supply. As they move through the enclosure, the carcasses rub against an electrode that is energised at the stimulating voltage. The current that flows through the carcass from the rubbing electrode to the conveyor makes the meat tenderer than it would otherwise be, a fact that is used as a selling point.

The physiological effects that lead to the perceived improvement in the quality of the meat will not be considered here. What is of relevance is the potential for serious electrical injury to anybody who enters the stimulator enclosure while the rubbing electrode is live. Given that suspended animals as large as cattle have to be moved into and out of the enclosure, it can be

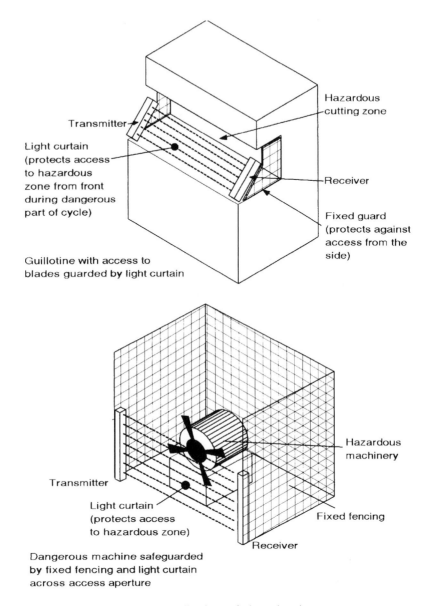

Fig. 13.16 Typical safeguarding applications of photoelectric systems.

appreciated that the entry and exit apertures are large enough for people to walk in and out with ease. The risk is enhanced by the fact that, as in all abattoirs, the environment tends to be wet. There are a number of ways in which the risk can be controlled but the most common is the use of light curtains to scan the floor across the entry and exit apertures. The light curtains are wired into the control system in such a way that the main con-

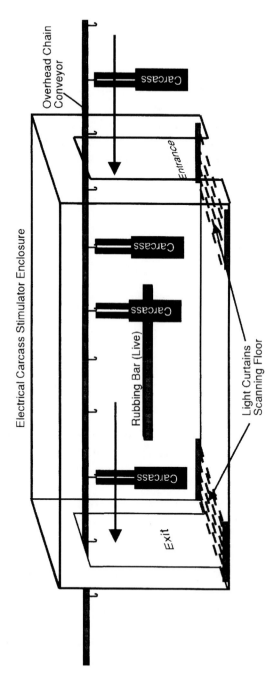

Fig. 13.17 Electrical meat stimulator with entry and exit apertures safeguarded using floor-scanning light curtains.

tactor providing electrical power to the stimulator is switched off if the light beams are interrupted. Some designs have light curtains that scan the whole of the floor area inside the stimulator to ensure that the device cannot be energised while somebody is inside the enclosure.

Photoelectric safety devices are 'safety components', as defined in the Supply of Machinery Safety Regulations 1992. A notified body, who will use BS EN 61496 as the baseline standard, must therefore check their conformity with the Regulations. The standard lays down general requirements for electrosensitive protective equipment, in Part 1, with Part 2 specifying particular requirements for photoelectric systems. For example, the standard specifies two types of device according to their fault tolerance, Type 2 and Type 4:

- Type 2 devices can be manufactured with a single output signal switching device but must have a periodic test to reveal any failures to danger. Such failures must be detected immediately, or as a result of the periodic test occurring after the failure occurs, or at the next actuation of the light curtain.
- Type 4 devices must be designed so that a single fault that adversely affects their detection capability will be detected and cause the device to 'lock-out'. Type 4 devices must have at least two output signal switching devices.

The next revision of the standard is likely to define a Type 3 device in which the fault tolerance characteristics fall between those of Type 2 and Type 4 devices.

The standard defines a series of optional extras. For example, there is an option for a stopping performance monitor, which allows for monitoring of the time taken for the machine to reach standstill after the beam has been interrupted. As another example, there is an option for external device monitoring which allows, for example, auxiliary contacts on contactors to be fed back to a light curtain controller and for the controller to enter a lock out condition if there were to be a detected disparity between the desired and actual condition of the contactor. The standard also specifies the optical requirements for the devices, including their immunity to optical interference, reflections and so on.

The interfacing arrangement between the electrosensitive protective equipment and the machinery control system is a particularly important facet of the safety system's design. It is the overall control system incorporating both the photoelectric guard and the machine's safety-related control system that determines the level of safety that can be achieved. Close attention therefore needs to be paid to the machine's control system and the way in which the photoelectric equipment's final switching contacts are connected into it. Guidance on this is published in HSE's Guidance Note HS(G)181 *Application of electro-sensitive protective equipment using light curtain and light beam devices to machinery.*

Another important issue is the separation between the light curtain and the hazardous parts protected by it. This is because the hazard must be removed by the time the light curtain and the machinery control system and any moving parts respond to the actuation of the light curtain by, say, an operator reaching through the light curtain. This matter is addressed in BS EN 999 *Safety of machinery – The positioning of protective equipment in respect of approach speeds of parts of the human body.* The standard specifies that the minimum distance, S, from the danger zone to the detection point shall be calculated according to the formula:

$$S = (K \times T) + C$$

where:
K is the approach speed of the part of the human body (e.g. 2000 mm/s for a hand/arm approach)
T is the overall system stopping performance
C is an additional distance in millimetres, based on intrusion towards the danger zone prior to actuation of the light curtain, which will typically be linked to the detection capability of the device.

In some applications, such as packaging machinery, products such as loaded pallets have to pass through a safety-related light curtain from, for example, the palletising machine into a despatch area. The light curtain guards the aperture in the machine's perimeter fencing to prevent humans entering the danger zones. This means that the light curtain must be muted, or effectively switched off, while the pallet load is passing through it, otherwise the pallet load would trip the machine's control system, and it must then be reinstated as soon as the load has passed through. The layout of the system should be such that the pallet load fills the aperture while it is passing through it, so as to prevent people passing through while the light curtain is muted. Sensors that detect the presence of the pallet load as it approaches the aperture normally activate the muting system. It is important that these sensors cannot be activated by anybody trying to defeat the light curtain.

Laser scanners

Another type of active optoelectronic device that is finding an increasing number of safety applications is the infrared laser scanner manufactured and supplied by companies such as Erwin Sick. This is a single device that uses a laser to scan the spatial volume in and around a hazardous zone over an arc of 180°. In Erwin Sick's PLS device, the maximum surveyed radius is 50 m, with two subsidiary sectors – one sector extends out to 4 m and is known as the safety zone, the other extends out to 15 m and is known as the early warning zone. The device is programmed by the users so as to ignore fixed objects in the surveyed area and to respond to objects, such as humans, that enter the two inner zones. An object entering the safety zone causes the device to output a signal that can be used to trip the operation of the hazardous machinery. If an object enters the early warning zone, an alarm signal is initiated. Objects down to 70 mm in size can be detected and resolved within the safety zone with a claimed response time of less than 80 ms. The devices are designed such that no single fault can lead to a loss of the safety function, making them suitable for medium risk safeguarding applications.

Emergency stops

Emergency stopping facilities must be provided where they will contribute to risk reduction. Therefore, they are not needed where they will not lessen the risk of injury or in hand-held portable and hand-guided machines. The particular requirements for the equipment are laid out in BS EN 418 Safety of machinery – Emergency stop equipment, functional aspects – Principles for design.

Emergency stops may be initiated by a number of different means including actuators such as mushroom-type push buttons, wires, ropes, bars, handles and footpedals that do not have protective covers. The devices must be designed such that they latch-in when actuated to generate a stop command, with the emergency stop command being maintained until the device is reset. The actuators must be coloured red and, where a background exists, it should be coloured yellow as far as practicable.

Most machines supplied nowadays have a mushroom-headed emergency stop button fitted on the control panel. On larger machines there will normally be more than one button, with all the buttons (or groups of buttons) wired in series. The pull-wire device is particularly suitable for conveyors and other long machines that have a hazard along their length. It consists of a long wire running alongside the machine and connected at its extremities to two limit switches connected in series in the control circuit, as depicted in Fig. 13.18. Pulling the wire operates one or both of the switches and causes the control circuit to respond, usually by deenergising a main contactor to switch off prime movers such as motors, hydraulic rams, pneumatic cylinders and so on. The device should latch in the off position and require a deliberate reset to allow power to be restored to the machine. Figure 13.19 shows the system redesigned so that the switches would operate

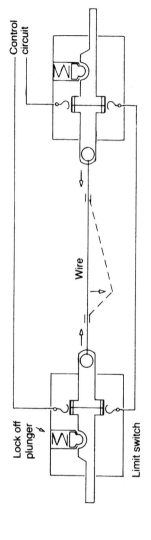

Fig. 13.18 Basic 'pull wire' emergency stopping system.

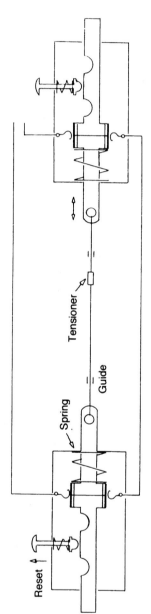

Fig. 13.19 Improved 'pull wire' emergency stopping system.

if the wire were to break. In this configuration the wire must be tensioned to ensure that both switches are 'on' at the same time.

Emergency stops must function as either a category 0 or category 1 stop, as defined in BS EN 60204-1. The former initiates immediate removal of power to the machine actuators and braking where necessary. The latter is a controlled stop with power being retained to achieve the stop and then removed when the stop has been achieved, such as the stopping of a paper-making line under the control of a PLC, with power being removed via contactors and similar devices when the high-inertia paper reels have been brought to a controlled stop. The common feature of the two categories is the eventual removal of the power source to the machine's prime movers.

Switches, push buttons etc.

Although the discussion so far has concentrated on devices that perform interlocking and trip functions, inputs to the control system processing unit are commonly in the form of signals from selection switches, push buttons and other devices. There is a huge range of such devices and, as far as safety is concerned, it will simply be noted that these devices are commonly used to implement start, stop and mode selection functions.

THE PROCESSING UNIT

As previously noted, the 'processing unit' lies at the heart of the control system, processing the signals received from the input devices and providing actuating signals to the output devices. In its very simplest form, it may be a single circuit, as shown in Fig. 13.20, that directly connects an input device, a switch, to an output device, a contactor coil, whose contacts switch a motor

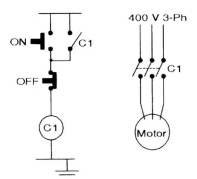

Fig. 13.20 Simple configuration.

on and off. It is more likely, however, that the processing unit will be somewhat more complex than this.

For many years, processing units in electrotechnical control systems were implemented using electromechanical devices such as relays, switches, push buttons and so on. This discrete component technology remains in widespread use today. One of the disadvantages of this type of technology, particularly in more complex applications, is that as the component count increases so does the amount of cabling needed to interconnect them all, with concomitant increases in cost and unreliability, and problems with maintainability.

Increasingly since the late 1970s, control systems have used programmable devices, principally programmable logic controllers (PLCs) and embedded microcontrollers, to implement control strategies in software. This type of technology provides designers with a considerably increased degree of flexibility and functionality, with increased processing speed and decreased wiring/component (and cost) overheads. PLCs can also be connected on to data buses to provide a network of systems capable of controlling large and complex manufacturing installations.

The ubiquitous PLC is essentially a microprocessor that takes inputs from a number of field devices and provides outputs to a number of output ports. The fundamental configurations are shown in Figs 13.21 and 13.22. Figure 13.21 shows a high-level schematic of a basic system structure. Figure 13.22 shows in a simplified way that the PLC scans data from a number of inputs, processes the data according to the program stored in its internal memory, and then outputs the processed data to a set of outputs. PLCs were, and still are in many cases, programmed using 'ladder logic' because it provided a

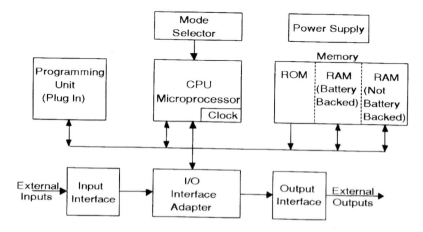

Fig. 13.21 Configuration of basic PLC.

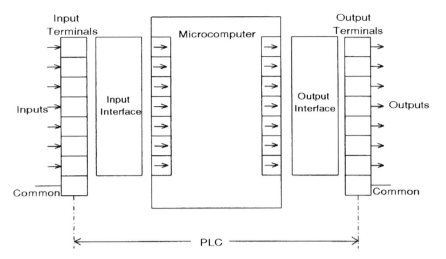

Fig. 13.22 Scanning data into and out of PLC.

representation analogous to the traditional hardwired systems. This eased the transition for many designers from discrete component technology to programmable technology. As an example of this, Fig. 13.23 shows how an element of a very simple hardware control system maps into the equivalent ladder logic program.

More modern devices are moving away from the use of ladder logic, largely because the complex functions that they offer are not amenable to the simplistic ladder logic representations. The languages are defined in an IEC standard, IEC 61131-3, which is the international standard for programmable controller programming languages. It specifies the syntax, semantics and display of five types of PLC programming languages:

- Ladder diagram (LD);
- Sequential Function Charts (SFC);
- Function Block Diagram (FBD);
- Structured Text (ST);
- Instruction List (IL).

One of the main benefits of the standard is that it allows multiple languages to be used within the same PLC. This allows the program developer to select the language best suited to each particular task. It also facilitates the production of modular and structured programs which, in turn, reduces errors and increases programming efficiency. These are significant advantages where software is used in safety applications.

Devices such as PLCs are used to control the operational functions of

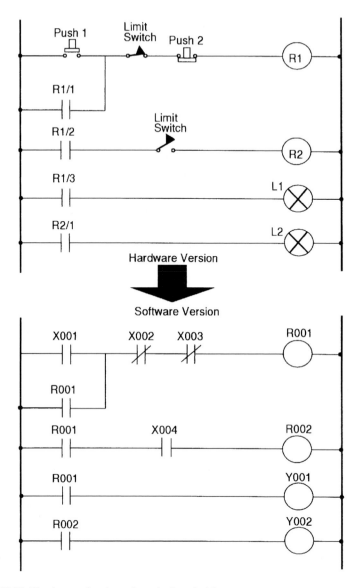

Fig. 13.23 Hardware circuit and equivalent ladder programme.

machines – to ensure that the machine does what it is meant to do. However, in the machinery safety field there has been a long-held concern about the integrity of programmable devices used to implement safety functions. The concerns have stemmed from the difficulty experienced in testing software with sufficient rigour to provide confidence that it is free of systematic faults – faults or 'bugs' designed into the software, both at the level of the operating system and at the application level. Whereas electromechanical control systems can have their reliability and failure modes predicted because there is a considerable amount of relevant data available, it is considerably less easy to predict the reliability and failure modes of programmable systems.

Another concern is the relative ease with which the software in PLCs and other programmable electronic devices can be altered. This raises the spectre of users of programmable systems that perform safety functions altering the code in a haphazard way without assessing the consequences on the safety performance of the system and, perhaps, losing control of the configuration management of the software. This latter point means that unauthorised or uncontrolled modifications, without the modifications being documented, can result quite quickly in the users not having an understanding of the software and how it is configured. Ultimately, this can lead to chaos and danger.

Imagine the situation, for example, in a 'fly-by-wire' passenger aircraft in which all the control surfaces (elevators, rudder, ailerons and so on) are controlled by computers which work to maintain the aircraft in steady and stable flight and which respond to inputs from the pilot and on-board sensors. It is unlikely that the flying public or the regulatory authorities would be too pleased if they felt that the safety of the aircraft was being compromised by engineers or aircrew making modifications to the software code in the computers to see what effect they would have on aircraft performance or to find out if they could improve fuel consumption. The need for tight configuration management and control of modification procedures in programmable systems that perform safety functions is of paramount importance. Whereas users in the high risk industries (nuclear, petrolchemical, aircraft and so on), where safety-related programmable systems are quite common, generally do have very good and comprehensive processes and procedures, the same is not true in the machinery sector.

These issues led the standards makers who wrote EN 60204-1 and EN 954-1 to insert notes in the standards to warn against the use of single channels of programmable electronics in safety applications. The net result of this has been a tendency, regarded by many as almost as a requirement, for safety functions to be hardwired to make them independent of the electronic and programmable electronic elements of control systems. Indeed, the standards stipulate that emergency stop systems must be hard-wired, and this has been interpreted as a firm requirement for many years now although, as will be explained later, advances in technology mean that such a restriction is becoming untenable.

An example of a design incorporating hardwired safety circuits is shown in Fig. 13.24. This shows how the outputs of the interlocking switch and the emergency stop actuator are routed not only to the input of the PLC

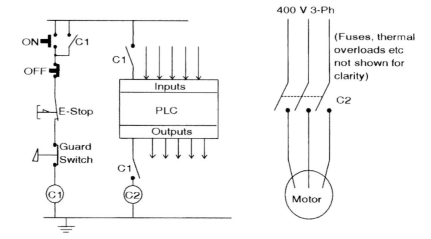

Fig. 13.24 Hardwiring the safety circuit.

controlling the machine's actuators, but also to the output side of the PLC, effectively bypassing the PLC. The aim of this configuration is to ensure that actuation of the interlocking or emergency stop devices will lead to the appropriate outputs being switched to the correct state by hardwired means, without reliance on the programmable elements of the control system. It also ensures that the programme correctly represents the current state of the machine. Although the diagram shows a contact being used to switch just one of the PLC's outputs, a more common configuration is for the contact to be placed in the power circuit to the PLC's output common module. This arrangement means that all the PLC's outputs would be switched off in the event of the contact opening.

In many modern applications, certainly those designed since the middle of the 1990s, the hardwired safety functions are routed through so-called safety relays manufactured and supplied by companies such as PILZ, Telemecanique and EJA. These are components that, in general, provide an inbuilt dual channel cross-monitoring facility aimed at providing a means of implementing high integrity control systems capable of meeting the fault tolerance requirements of EN 954-1; the concepts are described later in this chapter.

Whereas the hardwiring of safety functions remains the dominant technique for ensuring their safety integrity, the development of 'safety PLCs' and safety buses for networking safety components is beginning to drive machinery designers towards the use of programmable systems for safety applications. This trend is explained later in the chapter.

OUTPUT DEVICES

In most applications the ultimate aim of the processing unit is to cause the machine's prime movers – principally electric linear and rotary motors, braking systems, solenoid-operated bolts and plungers, and hydraulic and pneumatic rams – to move in the required direction at the appropriate speed and acceleration. This is normally achieved via interposing components that are directly switched and controlled by the signals output of the processing unit, although some devices are directly switched by the processing unit. The most common of these components are relays, contactors, solenoid-operated valves, and speed controllers.

In safety applications on machinery, the most common output reaction to a safety-related input such as an interlocked guard being opened or the emergency stop button being pushed, is for power to be removed and/or for brakes to be applied and for them to remain applied until a reset signal is provided. In electrical-only systems, the reaction is to have power disconnected to remove the risk of electrical injury.

Output devices used to implement safety functions must be adequately rated for their duty and have an appropriate level of reliability and fault tolerance. They should also be designed so that, wherever possible, faults are revealed and do not lead to the loss of the safety function; for example, electromechanical brakes should be designed to be held-off by being energised so that a loss of supply fault will lead to the brake being applied. Braking systems can be quite complex in practice, but their principles are quite straightforward. The main techniques are as follows.

Mechanical brakes

Mechanical brakes use friction linings to stop mechanical parts when the brake is applied. The brakes are normally held off under electrical, pneumatic or hydraulic power and are spring-applied.

Electrodynamic brakes

There is a variety of types of electrodynamic braking systems: d.c. injection, capacitive, plugging, and regeneration being the most common. Note that all of these types of braking system require current to flow for the system to function – they may therefore all fail to danger under certain fault conditions and this needs to be considered by carrying out a risk assessment during the design stages.

D.C. injection braking

In this type of system the a.c. electric motor is first disconnected from its supply and a d.c. current injected into its windings for sufficiently long to stop the motor. The circuit diagram in Fig. 13. 25 shows the basic features of this type of system. Pressing the 'start' button energises relay R and closes the direct-on-line starter contacts to supply three-phase power to the induction motor, which will then accelerate to its working speed. The starter is held on via contacts R1.

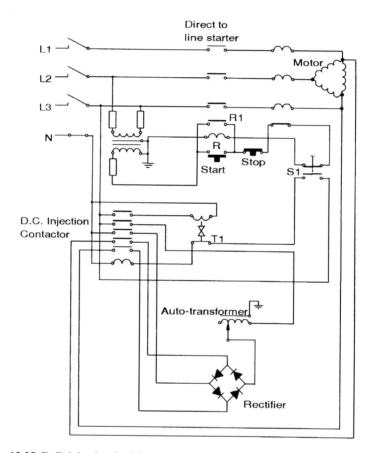

Fig. 13.25 D.C. injection braking system.

If the processing unit receives a safety signal such as an emergency stop it responds by toggling switch S1. This breaks the supply to relay R, tripping off the direct-on-line starter contacts and disconnecting the motor. Switch S1 also closes the circuit containing the coil for the d.c. injection contactor, closing the contactor. This provides a supply to the rectifier and leads to a d.c. current being injected into the coils of the motor, bringing the motor to a stop. After a period, the timer operates to break contacts T1, deenergising the d.c. injection contactor and removing the d.c. supply to the motor. The size and duration of the d.c. injection current must be set to the correct values for the system to work correctly.

Capacitor braking

In capacitor braking, the motor is disconnected from the supply, then connected to a capacitor for a preset time and then short circuited during the final deceleration stage to stop the motor.

Plugging

Plugging involves a reversing contactor that, when operated, swaps two of the phases on the supply to the motor – the motor is then said to be 'plugged'. This reverses the electromagnetic field that drives the motor and causes the motor to decelerate. If the plugging is not removed when the motor stops, the motor will then reverse its direction, which may be hazardous. Care must therefore be taken in the design of this type of braking system.

Regeneration braking

In regeneration braking, the power supply is removed from a motor, which then acts as a generator. This causes the motor to decelerate.

SAFETY INTEGRITY ISSUES

Thus far the components that make up a control system have been described without considering in detail how they are integrated to ensure that the systems have sufficient robustness, or safety integrity, for the safety application in which they will be used. Safety integrity refers to the reliability with which the system will perform the safety-related tasks it has been designed to perform, as well as the extent to which it will continue to perform its safety functions in the event of faults occurring. We are therefore interested in system performance in terms of reliability and fault tolerance.

It would seem, intuitively, that the greater the contribution that a control system makes to safety the more reliable it should be and the more resistant it should be to faults that would adversely affect the safety function. These are important concepts that are covered in two standards that are relevant to machinery safety applications and which are the subject of this section of the chapter: BS EN 954-1 and BS IEC 61508. There are many other national and international standards that cover system safety issues but which tend to be targeted at very specific applications, an example being defence and military standards (DEF-STANs and MIL-SPECs) covering avionic systems in aircraft. These standards are not considered in this book.

BS EN 954 : 1997 Safety of machinery – safety-related parts of control systems. Part 1. General principles for design

BS EN 954-1 : 1997 is the British version of EN 954-1, which was written by CEN Technical Committee 114, a European standards-making committee. It is a Type B1 standard which is meant to be used by technical committees preparing Type B2 and Type C European standards. It is interesting to observe that the UK voted against this standard at the formal vote stage of its development, for reasons that will become apparent.

The standard provides guidance on the design of safety-related parts of control systems on machinery. This would typically be guard interlocking circuits, stop and emergency stop circuits, and trip devices. The standard was written for the machinery sector, but it has broad application to safety-related systems although, as will be explained later, its use is generally restricted to 'non-complex' systems. Non-complex systems are usually regarded as those that employ devices, such as electromechanical components, whose failure modes can straightforwardly be determined.

The aim is to ensure that those parts of a control system that perform safety functions have adequate safety performance for the application. Safety performance, as previously noted for safety integrity, relates to the reliability and fault tolerance of the safety-related parts, and their ability to provide the required level of risk reduction. It is therefore axiomatic that the process of satisfying the requirements of the standard for any particular machine starts with a risk assessment of that machine.

In the hierarchy of measures to control risk, it is preferred that risks are designed out. Those that remain will then need to be safeguarded using a variety of risk reduction techniques such as the use of fixed guards. Some of the techniques may make use of electrotechnical control systems. The risk assessment should therefore establish the safety functions of the control systems which will then allow the designer to identify the safety-related parts and to specify their contribution to risk reduction. The extent of this contribution will influence the design measures that must be taken. In the very large majority of machinery applications, the risk assessment and measurement of risk reduction need only be a qualitative assessment – it is rarely the case that any form of quantified risk assessment would be justified on machinery. This means that a significant amount of professional judgement comes into the assessment of the contribution that safety-related parts make to risk reduction. BS EN 1050 provides guidance on how the risk assessment may be conducted.

The standard classifies the design measures into five categories – B, 1, 2, 3 and 4 – on the basis of reliability and tolerance to faults. The main features of these categories are as follows:

- *Category B.* The safety-related parts are to be designed, constructed, selected, assembled and combined in accordance with relevant standards so they can withstand the expected influence. The design is such that the occurrence of a fault can lead to the loss of the safety function.
- *Category 1.* The requirements of B apply, plus the use of well-tried components and well-tried safety principles. This type of system can suffer the loss of the safety function on the occurrence of a fault, but the event is less likely than for a category B system.
- *Category 2.* The requirements of category 1 apply, plus the safety function must be checked at suitable intervals by the control system. In this type of system the occurrence of a fault can lead to the loss of the safety function but the loss is detected at the next check.
- *Category 3.* The requirements of category 1 apply, plus the design must ensure that no single fault in the safety-related parts leads to the loss of the safety function, and the fault is detected whenever reasonably practicable. In this type of system, the safety function will be performed even when a single fault exists, and the fault may be detected, but might not be, and an accumulation of faults will lead to a dangerous situation.
- *Category 4.* The requirements of category 1 apply, plus the design must ensure that a single fault in the safety-related parts does not lead to the loss of the safety function, and the single fault is detected at or before the next demand upon the safety function. If this is not possible, then an accumulation of faults must not lead to the loss of the safety function. This means that the safety function will be performed in the event of faults occurring.

It should be noted that the categories are not a measure of the risk on the machine, as is commonly assumed. They are simply a means of classifying the required safety performance of the safety-related parts of the control system on a machine. Indeed, a single control system may contain subsystems that have differing categories and which are implemented in different technologies.

It is also important to note that the standard does not contain advice on the design techniques that will achieve these categories. However, most manufacturers of safety components publish literature that provides guidance on the system architectures that can be used to satisfy the requirements of the various categories. In general terms, category B and 1 control systems will employ single channel architectures while category 2, 3 and 4 control systems will use dual channel architectures with redundancy and, possibly, diversity. The fault detection requirements imply the use of cross monitoring between the redundant channels. In systems where sophisticated diagnostic capabilities exist for the detection of dangerous failures in single channel architectures, these generalisations may not apply.

The standard contains, at Annex B, a risk graph that can be used to determine the appropriate categories for a given application. The main features of the graph are illustrated in Fig. 13.26. Although the Annex and the risk graph are meant to be informative in nature, the graph has tended to become the centrepiece of the standard, mainly because it is seductively easy to use. In many circumstances, its ease of use has led to it being used as a general risk assessment tool to generate a global category for the machine, ignoring the need to consider each safety function and its individual contribution to risk reduction. The graph also gives the impression that the

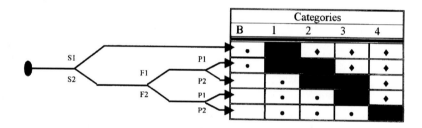

S: Severity of injury
S1:   Slight (bruising, fracture etc, reversible injuries)
S2:   Major (loss of limb, loss of sight, fatal and irreversible injuries)

F: Frequency and period of exposure
F1:   Infrequently exposed to hazard or exposure for a short period
F2:   Frequently exposed to hazard or exposure period is long

P: Possibility of avoiding the hazard
P1:   Possible
P2:   Less possible

- Preferred category in light of degree of risk reduction required

- Measure has higher integrity than justified by the required risk reduction

- Category may be used in conjunction with other measures

Fig. 13.26 Risk graph from EN9541-1 Annex B.

categories are meant to be hierarchical in nature, which they are not. The presence of the graph, and the foreseeability of its misuse, was one of the reasons that the UK voted against the standard.

What can be said about the categories is that:

■ Categories B and 1 are largely associated with the reliability of the constituent components.
■ Categories 2, 3 and 4 are largely associated with the structure of the system, because of their fault tolerance requirements.
■ For any single technology, categories 1, 2, 3 and 4 have a better safety performance than category B.
■ Categories B, 1 and 2 permit the loss of the safety function.
■ Categories 3 and 4 will not fail due to a single fault.
■ Category 4 has the best overall safety performance because an accumulation of faults is considered.

The UK's action of voting against the standard was a prescient move because the incorrect or inappropriate use of the risk graph has caused considerable problems with its interpretation, so much so that CEN took the unusual step of releasing a document – PD CR 954-1-100 : 1999 Safety of machinery, Safety-related parts of control systems, Guide on the use and application of EN 954-1 : 1996 – to explain how the standard should be used. The main difficulty was, and still is in many circumstances, that many machinery designers and users of machinery used the graph to carry out generic risk assessments, frequently concluding the exercise with meaningless statements such as 'this is a category 3 machine', or similar. This led them to conclude that their machine needed to have, or be upgraded to have, dual channel redundancy and fault detection in the safety-related control system when, in many circumstances, this level of safety performance was unnecessary.

Manufacturers of safety devices, especially those of 'safety relays', were quick to see the market potential, and the installation of safety relays in machine control systems soon became regarded by many as a fundamental requirement for compliance with the law. By way of explanation, the basic concept of these relays is depicted in Fig. 13.27. When power is first applied to the device, relay R3 is energised, which causes relays R1 and R2 to become energised. The output contacts of R1 and R2 close, and the output contacts of R3 end up in the closed position as relay R3 deenergises. If the emergency stop button is pressed, relays R1 and R2 deenergise, leading to the opening of the output contacts, thereby removing power from connected devices such as contactors. If the contacts of either relay R1 or R2 were to fail in the closed position, the output would still be switched off by the other relay's contacts. If this fault were to occur, the relay could not be reset because relay R3 could

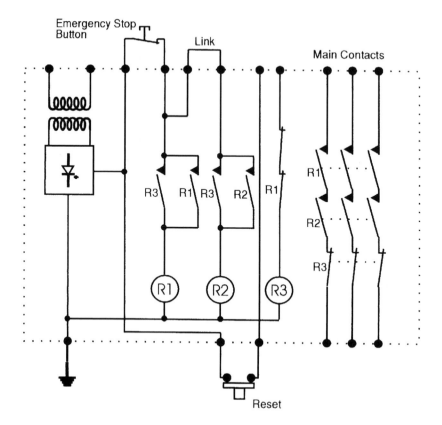

Fig. 13.27 Basic configuration of safety relay.

not be energised, thereby revealing the fault. This technique for achieving single fault tolerance and failure detection is known as dual channel cross monitoring. The technique has been in use for many years using discrete components – the safety relays encapsulate the function within a single component.

These safety relays do indeed provide a facility for fault-tolerant and monitored control circuits, which can be used to meet the requirements of category 2, 3 and 4 control systems. In that sense, they offer a significant contribution to the achievement of safe machinery control systems. An example of a category 3 system using a safety relay is shown in Fig. 13.28. This shows a dual channel architecture with two contactors controlling power to the motor, the contactors being switched by the main contacts in the safety relay. There are auxiliary contacts on the contactors that are incorporated in the reset circuit of the safety relay, thereby monitoring the contactors. Any single fault occurring on this system will not cause the loss of the safety function, the safety function being the disconnection of the motor

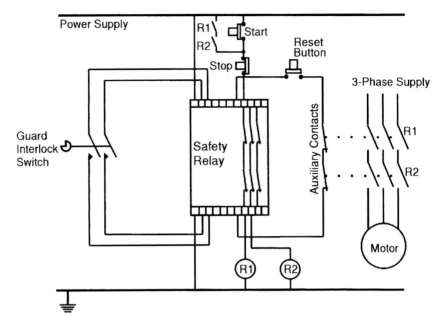

Fig. 13.28 Category 3 system using a 'safety relay'.

when the guard is open. Moreover, the single fault will be detected. This type of architecture is appropriate where the interlocking circuit is making a significant contribution to risk reduction. An example would be the inter-locked access door on a bottling machine through which an operator has to go once an hour to replace labels and pick up debris and where the hazard is loss of an arm if the machine were to move unexpectedly.

There are a large number of machinery safety applications in which there is no justification, in terms of risk reduction, for having a dual channel single-fault tolerant single-fault detection architecture such as this. For example, an interlocked access door on an electrical panel in which the components are all protected to IP2X standard and which is only accessed once a week would only warrant a single channel category 1 interlock.

In these low integrity (i.e. category B and 1) systems it is difficult to argue the case for using safety relays. Yet there are many examples where safety relays are used and where there is one interlocking or emergency stop circuit on the input side of the safety relay and only one contactor or interposing relay fed from the safety relay. This is depicted in Fig. 13.29. This is a category 1 configuration in so far as the system can fail to danger in the event of a single fault such as the contactor sticking in. It will be appreciated that a safety relay used in this type of configuration is not contributing to the reliability or fault tolerance performance of the circuit. Indeed, it could be

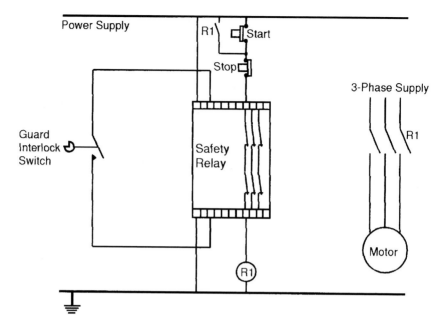

Fig. 13.29 Commonly used single channel category 1 system.

argued that its main contribution is an increase in the cost of the interlocking circuit and the possibility of decreasing the overall reliability of the system by virtue of having introduced more components than necessary.

The lesson from this is that, whereas BS EN 954-1 is an important and very useful standard, users of it must take full account of all its sections and not make the mistake of ignoring the text in favour of using the Annex B risk graph.

Before concluding on this standard, an observation needs to be made about its suitability for use in applications where programmable systems are being used to control safety functions. The problem here is the lack of information in the standard on the verification and validation of the software that would form part of a programmable system, such as a PLC, used in safety applications. The lack of such information makes it very difficult for the designer to tie the definitions of the standard's categories into the reliability and fault tolerance of the software, creating real difficulties in the use of the categories in a way that would be meaningful.

The writers of the standard appear to have recognised the problem and included a note advising that single channel programmable systems should be avoided in safety applications; this coincides with the advice contained in BS EN 60204-1. Indeed, the main text offers no advice on the detailed validation procedures for programmable systems, although Annex E refers to other publications for advice on the matter.

One conclusion from this is that, given that it is very difficult to predict the failure modes of complex programmable systems, specifying EN 954-1 categories for safety-related programmable systems borders on being meaningless. It is in this area that a new IEC standard, IEC 61508, offers some assistance.

IEC 61508 Functional safety of electrical/electronic/programmable electronic safety-related systems

The international standard IEC 61508 provides guidance on mechanisms that can be used to ensure that safety-related electrotechnical systems implemented in electrical, electronic and programmable electronic technologies are safe. It comprises seven parts, the first of which were finalised and published in 2000, some 15 years after the IEC initiated work on the drafting of the standard. The seven parts have the following titles:

- Part 1. General requirements
- Part 2. Requirements for electrical/electronic/programmable electronic systems
- Part 3. Software requirements
- Part 4. Definitions and abbreviations
- Part 5. Examples of methods for the determination of safety integrity levels
- Part 6. Guidelines on the application of IEC 61508 Part 2 and Part 3
- Part 7. Overview of techniques and measures

The standard is large and complex, and its contents are not easily absorbed. Whereas this may not be a particular issue for companies developing safety-related systems for the likes of major petrochemical companies, it may act as an impediment to its adoption by small and medium size machinery manufacturers. Moreover, it is generic in nature, meaning that it is not targeted at any particular applications, although the thrust of it is more appropriate for complex safety-related control systems in the process, nuclear, railway and similar industries than for simple non-complex machinery control systems.

However, in the context of the discussion in this chapter, the standard has the significant advantage over EN 954-1 that it describes how the functional safety of complex electronic and programmable electronic control systems can be assured. This means that machinery manufacturers now have available a formal standard that they can use to assure the safety of PLC and other programmable controllers used in machinery safety-related systems. In recognition of this, CEN is in the process of trying to map the principles of the standard into a new machinery sector version of IEC 61508, currently identified as prEN 62016. It is yet to be seen how this new standard, when it is eventually published, will relate to EN 954-1.

At IEC 61508's core is the definition and explanation of a safety life cycle. The overall safety life cycle for safety-related control systems is depicted in Fig. 13.30. The important attribute of this concept is that it imposes a formal structure on the planning and management of the specification, development, installation, use, maintenance, modification and final disposal phases of the life of safety-related systems.

The specification of safety requirements is an essential step, with the requirements based on the contribution that the system is making to risk reduction, taking into account the contributions made by other technologies such as pressure relief valves and any other external risk reduction measures. The concept is fundamentally the same as that set out in EN 954-1, although the realisation of it is entirely different.

The specification will lead to the identification of safety functions, again as recommended in EN 954-1. IEC 61508's safety functions are defined in terms of their functionality and their safety integrity. The safety integrity of each safety function is roughly equivalent to the concept of the categories of EN 954-1. However, whereas the latter's categories are loosely defined in terms of their fault tolerance capability, IEC 61508's safety integrity levels are much more tightly defined in terms of their target failure measure. There are four levels of safety integrity, ranging from safety integrity level (SIL) 1 to SIL 4, with SIL 4 being the most safe. There are two types of safety integrity level: for safety functions that operate in low demand mode of operation, and for safety functions that operate in high demand or continuous mode of operation. An example of the former would be an emergency stop facility on a machine. An example of the latter would be a temperature sensor monitoring a safety-critical process. The quantitative values of the SILs are shown in Tables 13.1 and 13.2.

Although the numerical values of the SILs can be tied into a quantitative risk analysis, the standard does allow the appropriate SIL for a safety function to be determined by qualitative means using risk graph techniques. Unlike the case of the risk graph in EN 954-1, IEC 61508 does provide guidance on the use of the how the risk graph technique should be implemented. Once a safety function has an assigned SIL, the designer needs to decide on the architecture, diagnostic coverage, proof test interval and other criteria that need to be adopted to ensure that the system achieves the specified SIL. The standard provides comprehensive guidance on all these techniques.

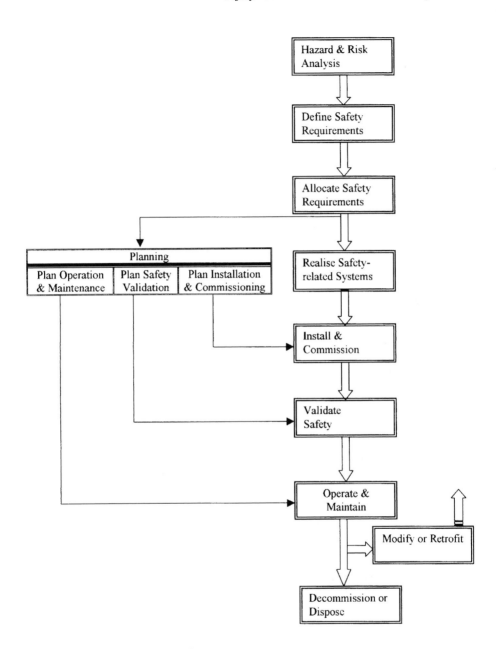

Fig. 13.30 IEC 61508 safety life cycle.

Table 13.1 Safety integrity levels – low demand mode.

Safety Integrity Level	Target failure measure (Average probability of failure to perform its design function on demand)
1	$\geq 10^{-2}$ to $< 10^{-1}$
2	$\geq 10^{-3}$ to $< 10^{-2}$
3	$\geq 10^{-4}$ to $< 10^{-3}$
4	$\geq 10^{-5}$ to $< 10^{-4}$

Table 13.2 Safety integrity levels – high demand/continuous mode.

Safety Integrity Level	Target failure measure (Probability of dangerous failure per hour)
1	$\geq 10^{-6}$ to $< 10^{-5}$
2	$\geq 10^{-7}$ to $< 10^{-6}$
3	$\geq 10^{-8}$ to $< 10^{-7}$
4	$\geq 10^{-9}$ to $< 10^{-8}$

In the context of programmable systems, the standard recognises the problem of systematic errors occurring in the hardware and software of complex systems, and the difficulties associated with estimating the probability of these faults occurring. To overcome this problem, the standard provides guidance on the measures and techniques that may be adopted during the design stage to avoid the introduction of systematic errors. The specific measures and techniques are beyond the scope of this book, although readers who are interested to know more about IEC 61508 will find a very useful and informative description of it in the special feature articles in the IEE *Computing and Control Engineering Journal*, Volume 11, Number 1, February 2000.

It has to be said that the standard, whilst an important basic safety standard, leaves open many questions that are now the subject of intense research and analysis. For example, when determining the safety integrity level of existing programmable systems, what credit can be taken if there is a large population of system in use and no systematic faults have been evident in its use? How should systems be treated when they have no direct safety function but their failure may have an impact on a safety-related system? How should human factors be considered when determining the safety integrity level of a system that has a human operator in the process loop? What is the safety integrity of a human being? And so on.

Another important follow-up to the publication of IEC 61508 is the development of conformity assessment schemes for safety-related systems based on IEC 61508. Such schemes have a wide range of objectives, many of which relate to improving competitiveness of industry, but one of which is to facilitate the development and use of safety-related systems based on modern technology. The UK's strategic initiative in this area, supported by the DTI, is known as the CASS (conformity assessment of safety-related systems) project and is being delivered by The CASS Scheme Ltd, a company limited by guarantee. The scheme, which is also covered in Chapter 14, will lead to the assessment of components, systems and organisations' functional safety capabilities against the requirements of IEC 61508, CASS assessment services being offered to industry by UKAS accredited certification bodies. The existence of such a scheme illustrates the importance of IEC 61508 and the impact it is already having in the field of safety-related systems.

TECHNOLOGICAL TRENDS

As has previously been alluded to, there are two technological trends that will have a significant impact in the design of safety-related systems in the machinery safety sector. These are the introduction of PLCs for use in safety applications, so-called safety PLCs, and the development of 'safety' field-buses. The availability of these systems will allow machinery designers to make full use of the advantages that programmable controllers and data communication networks have over the traditional hardwired safety solutions.

Safety PLCs

Safety PLCs are manufactured by companies such as PILZ and Siemens. In general terms, they achieve high levels of hardware safety integrity by incorporating into a single PLC package two or three separate controllers with voting strategies being implemented on their outputs.

In the PILZ PSS safety PLC, for example, there are three separate controllers with a three out of three voting system at their outputs. The controllers operate in parallel, processing input data independently – the overall output will only be enabled if the output from each controller is the same. The controllers in the PLC are from different manufacturers and have different operating systems, providing a considerable degree of diversity and redundancy. In this configuration, the standard non-safety-related programmes are run on just one of the three controllers, whereas safety-related software is implemented in the triple-voting section.

Siemens, on the other hand, adopts a slightly different philosophy in its safety PLCs. In its SIMATIC S5 range of safety PLCs, for example, there are two processor units linked by a fibre optic cable over which data is transferred to allow data comparisons to be carried out. If disparities are detected, the system enters a stop condition.

The hardware safety integrity of these PLCs has been certified to category 4 of EN 954-1 and, in the case of the Siemens units, to SIL 3 of IEC 61508.

Although the safety integrity of the hardware elements of these devices appears to be well assured by the manufacturers, there is always the question about the integrity of the software written for particular safeguarding applications. Can it reasonably be expected that engineers in end users' premises will be able to produce software that matches the safety integrity of the safety PLC's hardware? It certainly seems that the manufacturers recognise this problem. PILZ, for example, provides pre-coded function blocks for the most common safety functions – emergency stop, gate monitoring and so on – as well as specialist modules for functions such as safety systems on hydraulic presses and burner management systems. Nonetheless, this is an area in which considerable care must be taken to ensure that the overall safety of the system is not compromised by poor programming standards and practices.

It can be anticipated that these devices will become increasingly common in a wide range of applications. The challenge for the system integrator is to ensure that designers do not become seduced into believing that the presence of a safety PLC is indicative of inherent safety. The input and output devices, sensors and actuators will continue to play a pivotal role in the overall safety integrity of a system, and their contribution, together with that of the software in the PLC, would be ignored at the designer's peril.

Safety buses

Digital data communication systems have existed for nearly 40 years, ranging from straightforward serial communications represented by the RS 232 standard common on personal computer serial output ports, to more modern local area networks such as Ethernet. In recent years, much effort has been put into the development of fieldbus systems that allow controllers, sensors and actuators to be implemented in modular and distributed formats. Whereas the developments in fieldbus systems have been mainly targeted at the process industries, the devices are finding increasing application in the machinery sector.

The physical medium used for fieldbuses is quite simple – it can be a shielded twisted pair cable or coaxial cable. The important characteristics tend to be in the data structures and protocols used, and these generally conform to published standards.

Two well-known fieldbus systems are Profibus and AS-I (Actuator Sensor

Interface). The former tends to be used where high levels of functionality and data rates are needed whereas the latter, which is based on the controller area network (CAN) protocol, is used in applications where there are lower functionality and simple input/output requirements. The manufacturers of these fieldbus systems have worked on developing them for use in safety applications, mainly to incorporate appropriate levels of fault tolerance or safety integrity. This has led to the availability of the Profisafe and AS-Isafe fieldbuses. In addition, PILZ has developed the SafetyBUS fieldbus for safety applications, which is again based on the CAN protocol, and the Open Devicenet Vendors Association has developed a safety version of the DeviceNet fieldbus called DeviceNet Safety.

The general aim of these developments has been to allow safety applications to access standard fieldbuses using safety-related communications protocols, without having to invest in too much extra separate hardware and software. Having said that, the SafetyBUS fieldbus is separate from the general plant or machinery fieldbus and is used to connect safety devices such as light curtains and interlocking switches to the safety PLC controlling the process. As an example, Fig. 13.31 shows how a safety PLC is connected to a fieldbus, such as SafetyBUS. The bus loops into and out of input/output modules into which are connected safety devices. The advantages of this type of configuration, in terms of flexibility and cost reduction when compared with the more traditional hardwired topologies, will be obvious.

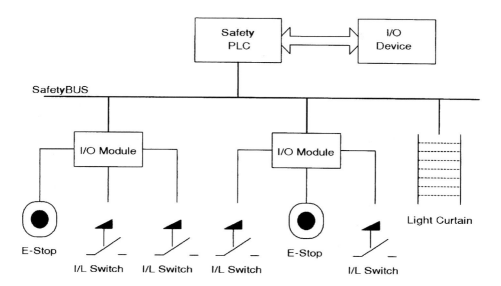

Fig. 13.31 Typical SafetyBUS architecture.

Chapter 14
Competence

INTRODUCTION

Many accidents are caused by people making mistakes when working on electrical systems, with the mistakes often being directly attributable to weaknesses in the competencies they needed to carry out their work activities safely. In the jargon, these accidents are known as failures in the systems of work. In addition to them, system technical failures, be they in power systems or in control systems, are frequently the result of poor design features, again often resulting from lack of competence on the part of the engineers who specified, designed and produced the systems. It is clear, therefore, that personal competencies play a particularly important role in the achievement of high standards of electrical and control system safety.

It is important that managers who have responsibility for ensuring the competence of engineers and technicians understand the factors that determine whether or not an individual is competent to perform particular tasks. For many people in this position it is tempting simply to equate competence with training and qualifications, working on the basis that individuals only require appropriate technical qualifications and some task-specific training to enable them to carry out specified tasks safely. Whereas training is undoubtedly an essential component of competence, it is not the only one. Furthermore, not all qualified people are competent, and lack of qualifications does not automatically mean that a person is not competent. On the other hand, the holding of a particular qualification may be one way of demonstrating competence. For example, an approved electrician is able legitimately to claim competence in the installation of electrical systems covered by BS 7671, by virtue of having completed a $4\frac{1}{2}$ year apprenticeship, passed the appropriate tests, and gained two years of further relevant experience.

In the safety arena, for a person to be competent to undertake work safely he or she should have a mix of sufficient and suitable training, experience, knowledge and other personal qualities. The 'other personal qualities' are largely associated with the individual's approach to work, with factors such as powers of concentration, sense of personal responsibility, diligence, integrity and maturity being important attributes. As an example of this, it is highly unlikely that a well qualified and experienced engineer could be competent to work on high voltage systems and make them safe for others to work on if he has a tendency to be a bit slapdash in his approach and to cut corners – such behaviour is a recipe for eventual disaster.

The issue of competence has been a concern to many organisations over the years and this has resulted in the development of competence-based qualifications that, increasingly, incorporate units on health and safety. For example, free standing health and safety units are being incorporated into vocational qualification standards set by national training organisations such as National Electrotechnical Training, the body that sets many of the standards for electrical vocational qualifications. This has led to the holding of appropriate National and Scottish Vocational Qualifications being one way of demonstrating basic skills and knowledge both in technical topics and in employment-specific health and safety issues.

In this chapter, the topic will be examined in the context of both electrical safety and the safety of electrotechnical control systems employing electrical, electronic and programmable electronic safety-related control systems. In addition, the increasingly common practice of using conformity assessment to demonstrate competence will be considered.

LEGAL ASPECTS

There is no definition of competence in health and safety legislation, although in the Management of Health and Safety at Work Regulations 1992, Regulation 6(5) says that 'A person shall be regarded as competent ... where he has sufficient training and experience or knowledge and other qualifications...' There is no shortage of other legislation that makes reference to the need for people to have the necessary safety-related skills and knowledge for their work, and the HSC/HSE certainly puts considerable emphasis on competence issues in its guidance material and approved codes of practice.

Employers have a duty to ensure that those working for them have the necessary safety-related competencies for the work being done. This is enshrined in the Health and Safety at Work Act, section 2(c), which requires employers to provide such information, instruction, training and supervision as is necessary to ensure, so far as is reasonably practicable, the health and safety at work of their employees. It should be noted that this duty does not just extend towards their direct employees – employers must also ensure that people not in their employment whose work activity may put their employees at risk, are given information and instruction. For example, an employer must make sure that contractors brought on to his premises to carry out electrical work are competent to do the work in a manner that does not put people on the premises at risk of injury. Of course, the contractor's managers have similar duties to their own employees.

These are the general legal requirements relating to the need to ensure that persons are competent. The Electricity at Work Regulations 1989 provide statutory requirements relating specifically to electricity by stipulating that persons must be competent to avoid electrical danger. It was noted in Chapter 6 that both Regulation 4(3) and Regulation 16 address the matter. Regulation 16 says:

> 'No person shall be engaged in any work activity where technical knowledge or experience is necessary to prevent danger or, where appropriate, injury, unless he possesses such knowledge or experience, or is under such degree of supervision as may be appropriate having regard to the nature of the work.'

The Memorandum of Guidance to the Regulations offers some help in interpreting these Regulations by defining the scope of 'technical knowledge or experience' as including:

- adequate knowledge of electricity;
- adequate experience of electrical work;
- adequate understanding of the system to be worked on and practical experience of that class of system;
- understanding of the hazards which may arise during the work and the precautions which need to be taken;
- ability to recognise at all times whether it is safe for work to continue.

These are very useful tests that can assist in determining the level of competence of individuals for conducting electrical work. Of course, the law also recognises that where persons do not have the appropriate competencies they must be under sufficient supervision by persons who are competent so as to ensure that the work is carried out safely.

When considering the legal duties on those who design, manufacture and supply electrical equipment, including equipment that incorporates safety-related control systems, the main legal duty is the Health and Safety at Work Act, section 6. This requires that articles for use at work (which includes electrical equipment) are designed and constructed so as to be safe and without risks to health when being set, used, cleaned or maintained by a person at work, so far as is reasonably practicable. Although this law does not specifically say so, it is clearly the responsibility of the suppliers to ensure that their design engineers have the appropriate competencies to ensure that these legal obligations are satisfied. A similar comment applies to other 'supply-side' legislation such as the Supply of Machinery (Safety) Regulations 1992, the Electrical Equipment (Safety) Regulations 1994, and the Electromagnetic Compatibility Regulations 1992.

ELECTRICAL COMPETENCE

Electrical competence is no different to other competencies in that it is gained from a mixture of qualifications (equating to knowledge), training, experience and personal qualities. Whether or not a person is competent to carry out electrical work safely will depend on the individual having the appropriate mix of formal qualifications, training and experience in relation to the complexities of the systems being worked on, the nature of the work, and the degree of risk. For example, an electrician who is skilled at installing cable systems but who does not energise or test them may not have the necessary competencies for carrying out the higher risk activities of live testing and fault finding on the systems once they have been energised, despite the fact that he may hold formal qualifications as an electrician. He or she will almost certainly need to undergo specialist training to gain the competencies needed to ensure safety during the live working activity.

In similar vein, the task of isolating and proving dead a low voltage circuit to allow work to be done on it requires different skills and knowledge when compared with the higher risk task of isolating and earthing a high voltage circuit forming part of a high voltage distribution network. Given these different levels of risk, the employer has the responsibility for ensuring that people exposed to them are sufficiently competent to manage the risk to themselves, and to others affected by their work, down to an acceptable level.

A common and well-proven technique adopted by many companies employing personnel to carry out electrical work, particularly where there are high voltage distribution systems on their premises, is to define in their electrical safety rules a range of electrical competencies and authorisations. This may include definitions covering, for example, competent person (electrical), authorised person (low voltage), authorised person (high voltage), and senior authorised person. People who are required to carry out electrical work are then allocated to one or more of these groupings according to an assessment of their individual competencies.

A competent person (electrical) would be a person whom the company has assessed to have the required competencies for carrying out work safely on its electrical systems. These competencies will usually stem from the individual having a formal qualification such as an NC or N/SVQ in electrical installation work, a degree in electrical engineering, a City and Guilds qualification in electrical maintenance, and so on. The person will need to be trained on the particular characteristics of, and risks associated with, the electrical systems on which he will be working, as well as on the systems of work implemented by the company. He will need to become familiar with the formal risk assessments that should have been carried out, and which should detail the measures to be taken to control the risks. He will also, in many cases, require to work under the supervision of a more experienced person for a period of time during which his competence will be assessed before being classed as a competent person (electrical).

An authorised person (low voltage) would be a competent person (electrical) who has the required competencies for carrying out work on the low voltage systems in the premises. The person should be authorised in writing, with a clear definition of the nature of the work he or she is authorised to carry out. This would include activities such as isolating low voltage circuits and equipment and, after a risk assessment has been completed, carrying out fault finding and testing work on systems that may be live.

On high voltage systems, the higher level of risk justifies a structured and formal safety management system covering the competence of the persons involved and the procedures to be adopted, which should usually be set out in high voltage safety rules. Competent persons (electrical) who are authorised to work on high voltage systems must receive special-to-type training in high voltage operational safety and on the high voltage equipment itself. The company should allocate responsibility to an electrical engineer, either from their staff or contracted in to do the job, to assess the competence of engineers aspiring to be authorised persons (high voltage). The assessment will check their general skills and knowledge, as well as their understanding of the electrical systems on which they will be working. The extent of their authorisations should be defined in writing, paying attention to the extent to which they are authorised to make systems safe and to issue electrical safety documents such as permits to work and sanctions for test.

A senior authorised person is normally the person, or one of a small number of persons, who is in charge of the electrical system and responsible for ensuring the safety of the system and of those responsible for working on it. A senior authorised person will normally be a highly trained and experienced individual with a particularly good knowledge and understanding of the electrical safety procedures at the plant.

Obviously authorised persons are in positions of considerable responsibility – they make systems safe for others to work on them and they therefore need to be conscientious and thorough, and these traits are essential elements of their overall competence. Any mistakes can have serious consequences, especially where high voltage work is concerned. There has been a small number of instances in which authorised persons who issued inappropriate electrical permits to work or failed to manage the electrical safety of the systems under their control, have been prosecuted following accidents to people working under the terms of the permits.

Employers should always keep in mind that competencies can be lost. A common example of this occurs in high voltage work. An engineer may attend high voltage operational training and, and after receiving local training, become authorised to operate high voltage switchgear. However, he or she may only have to carry out, on average, one high voltage switching operation per year, or even less frequently. In these circumstances not much operational experience is gained in the work and, over time, the procedures may be forgotten and competence lost. One way of guarding against this is for the person to attend refresher training on, typically, a five-yearly cycle.

The problem of checking the competence of electrical contractors has already been alluded to, but what techniques are available to ensure that enough is done to ensure that a selected contractor is competent? One option is to ensure that the companies being considered at the tendering stage are members of a recognised trade association such as the Electrical Contractors' Association (in England and Wales) or SELECT (in Scotland), or are members of the National Inspection Council for Electrical Installation Contracting (NICEIC). The NICEIC is a non-profit making body and a registered charity that maintains a roll of approved contractors that meet its enrolment rules and national technical safety standards, including BS 7671. Inspecting engineers employed by these bodies make annual visits to member companies to assess their technical capability and to inspect samples of their work, so membership gives some confidence in the capabilities of the companies.

In addition to membership of these bodies, checks should be made of the companies' electrical safety rules, to see that these exist and that they are relevant to the planned work. The companies' risk assessments and method statements should also be checked, to ensure that a risk-based approach to work planning is being adopted. Evidence should be sought of the competence of the individual tradesmen who would be working on the premises, as well as evidence of previous work done satisfactorily by the companies in the same field as the planned work. These types of checks, if conducted conscientiously and thoroughly, would be a reasonable approach to the task of selecting competent electrical contractors.

SAFETY-RELATED CONTROL SYSTEMS

The topic of safety-related electrotechnical control systems is covered in Chapter 13 where it is noted that such systems can comprise a mixture of electromechanical, electronic and programmable electronic equipment and can have designs that range from being quite simple to being extremely complex. In many cases, the systems can have safety functions or protection functions in which the consequences of failure can be severe. Safety-related control system failures leading to an explosion in a petrochemical site, or a collision between aircraft being controlled by an air traffic control system, or the loss of the protection systems in a nuclear power station, are examples in which the potential loss of life and environmental damage would generally be regarded as unacceptable.

This places much emphasis on the engineering of sufficient safety integrity into the safety-related systems to ensure that risks affected by those systems are reduced to a level that is as low as reasonably practicable (ALARP). There are many methods and techniques for achieving this. The IEC standard 61508 Functional safety of electrical/electronic/programmable electronic safety-related systems, for example, advocates a risk-based approach to determining the required safety performance of safety-related systems, and does so in the context of an overall safety life cycle. The success in the implementation of such an approach demands a suitable level of competence in the engineers involved in the determination of the safety requirements and in the realisation of such systems.

There has not been much guidance available on the competencies required for working in this area. This was a deficiency that needed to be remedied as the progressive introduction of programmable elements into systems that performed safety functions, such as PLCs used in machinery interlocking circuits, resulted in the systems becoming more complex, and increasingly difficult to assure for safety.

During the late 1990s this matter was addressed by the Institution of Electrical Engineers (IEE) and the British Computer Society (BCS), many of whose members work in industry on the development of safety-related systems. They studied the matter in collaboration with the Health and Safety Executive, which has played a leading role in producing guidance in this area and whose Technology Division's engineering inspectors were actively involved in the production of IEC 61508. The result of this collaboration was an IEE guidance document published in 1999, Competency guidelines for safety-related system practitioners, which is intended to form the foundation of continuing professional development programmes for engineers working in the field. It is a piece of work which industry is well-advised to adopt, if only to be able to demonstrate to potential clients that its engineers have the competencies needed to meet the increasingly demanding requirements for complex, but safe, control systems.

The document identifies the following 12 functional areas in which competence is required to support the specification, development, procurement, operation and maintenance of safety-related control systems:

- corporate functional safety management;
- project safety assurance management;
- safety-related system maintenance;
- safety-related system procurement;
- independent safety assessment;
- safety hazard and risk analysis;
- safety requirements specification;
- safety validation;
- safety-related system architecture design;
- safety-related system hardware realisation;
- safety-related software realisation;
- human factors safety engineering.

For each of these functions, the document provides an outline of the key responsibilities and describes the functional, as well as task-related, competencies. It does this for three levels of competence; in increasing levels of competence these are the supervised practitioner, the practitioner, and the expert. The document also provides guidance on how a competence scheme can be operated in these functional areas, how competence can be assessed, and how the outcome of the competence assessment can be recorded.

The difficulty that many people face is that documents such as this tend to be difficult to read and assimilate. Despite this, it is advisable to persevere and to put in place the recommended procedures and practices, or procedures and practices that will achieve the same goals in the management of personal competencies. The benefits far outweigh the problems.

CONFORMITY ASSESSMENT

One consequence of the increasing realisation that competence has a direct impact on health and safety has been the development of conformity assessment schemes covering competence. Conformity assessment, in its generality, is an activity concerned with determining directly or indirectly that relevant requirements have been met, and is most commonly used to determine if a product, system, process or a person's competence meets a defined specification. In the context of this chapter, conformity assessment relates to the process of determining whether people working in the electrical or safety-related control systems sectors can perform to the required level of competence.

Organisations that carry out this type of assessment work are known as certifying bodies and there are mechanisms in place whereby these bodies can be accredited as meeting appropriate standards, usually those in the EN 45000 series. The sole Government–approved accreditation body in the UK is the United Kingdom Accreditation Service (UKAS), which is a non-profit-distributing private company limited by guarantee.

An example of an accredited certifying body in the electrical sector is the NICEIC, which has been assessed by UKAS as meeting the requirements of EN 45011, the European standard for the certification of products, processes and services. This means that the NICEIC is able to issue accredited enrolment certificates to its approved contractors. This will include the assessment of the competence of electrical contractors under a proposed scheme known as the electrotechnical assessment scheme which is being developed to enable contractors to register as 'competent enterprises'. Such enterprises will, should the scheme eventually proceed, be recognised as competent to self-certify their work under changes that are being proposed to the Building Regulations.

Turning to the field of safety-related control systems, a significant initiative in the UK is the development of a scheme known as the conformity assessment of safety-related systems (CASS), already mentioned in Chapter 13. The initiative, which is supported by the DTI under its sector challenge programme and which will be championed and implemented by a UKAS-accredited company known as CASS Scheme Ltd, is concerned with conformity assessment for systems based on the aforementioned standard IEC 61508. The scheme has a broad scope and covers all those involved in the design, development, manufacture, implementation, support and application of complete safety-related systems and their components.

The CASS scheme is mainly concerned with allowing companies who carry out these functions to have their products and processes certified as conforming with the requirements of IEC 61508. This will bolster their position in the safety systems marketplace and will also facilitate developments in, and confidence in the use of, implementation of the standard. These are largely trade issues, but they will have an important contribution to improvements in the functional safety of safety-related systems and to being able to demonstrate that safety-related systems are fit for purpose. The competence of those who are designing, developing, supporting and applying safety-related systems will be a significant factor in the assessment process.

As a final observation, it is clear that these conformity assessment schemes have their roots in trade issues rather than in health and safety, but there are undeniable benefits in the health and safety field. For example, certification and registration schemes in the building industries, including the electrical sector, are favoured as one means of countering the 'cowboys' who charge exorbitant fees for minor works which are frequently of dubious quality and/or are unsafe. Registering competent practitioners will allow clients to select contractors with demonstrably adequate competencies and business practices, but it will also provide some assurance about their health and safety practices and about the safety of any electrical work they do.

Chapter 15

Electrical Equipment in Flammable and Explosive Atmospheres

INTRODUCTION

The potential for electrical equipment installed in flammable and explosive atmospheres to act as an ignition source has long been recognised. Arcs and sparks generated in normal operation or under fault conditions, and high surface temperatures on equipment enclosures, can ignite such atmospheres, causing explosions or fires that have the potential to cause serious injury and considerable economic loss. In addition, electrostatic discharges can act as ignition sources.

Over the years a considerable amount of knowledge has developed on the means by which the risks can be controlled, so much so that serious incidents arising from electrical ignition are now infrequent. The purpose of this chapter is to explain the principles of the main techniques. The first section covers flammable gas/air mixtures, and later the risks associated with explosions in dust-laden atmospheres are considered. The chapter does not cover the broader issues associated with controlling risk from flammable and explosive atmospheres, such as substituting for non-flammable materials where possible and containing the flammable materials to avoid the formation of an explosive atmosphere; the intention is to explain what needs to be done to control the risk of ignition from electrical equipment.

CHARACTERISTICS OF FLAMMABLE MATERIALS

A flammable atmosphere is associated with the presence in air of an appropriate concentration of flammable gases and vapours. For ignition to occur, there must be an ignition source with sufficient energy to ignite the particular gas/air mixture. The combination of gas, oxygen and ignition source is commonly referred to as the 'fire triangle'.

A flammable liquid is defined, in the Highly Flammable Liquids and Liquefied Petroleum Gases Regulations, as one that gives off sufficient vapour at a temperature of less than 32°C as to be capable of ignition and to support combustion, i.e. once ignited it will continue to burn when the source of ignition is removed. The lowest temperature at which a liquid will produce flammable vapours is known as the flashpoint. The ignitable concentration of any flammable material in air lies between the upper and lower explosive limits (UEL and LEL). A rich or lean mixture outside these limits will not ignite. The ignition temperature is the lowest temperature at which the material will ignite due to the application of heat.

Some liquids which do not emit sufficient vapour at normal ambient temperatures to be classed as flammable are ignitable when sprayed in a fine mist; Avtur, the paraffin used in jet engines, is an example.

HAZARDOUS AREAS

Where there are flammable materials that constitute an explosion risk, the locations in which the explosion risk exists are called hazardous areas. These areas are classified into zones according to the extent of the risk, using guidance published in BS EN 60079-10 : 1996 Electrical apparatus for explosive gas atmospheres, Part 10 Classification of hazardous areas. The zones are also defined in the European Directive on minimum requirements for improving the safety and health protection of workers potentially at risk from explosive atmospheres. The zones are:

- *Zone 0:* areas where an explosive gas atmosphere is always present or is present for long periods, such as inside reaction and mixing vessels or storage tanks.
- *Zone 1:* areas where an explosive gas atmosphere is likely to occur during normal operation.
- *Zone 2:* area in which an explosive gas atmosphere is not likely to occur in normal operation, and if it occurs it will do so infrequently and will exist only for a short time.

Employers in control of hazardous areas need to determine the locations and extents of these zones in each of the hazardous areas, an exercise known as hazardous area classification. At present, in 2001, there is no specific legal duty to carry out hazardous area classification, apart from the general duties to carry out risk assessments and, in the context of mines, in Regulation 19 of the Electricity at Work Regulations. However, the Protection of Workers Potentially at Risk from Explosive Atmospheres Regulations expected to be enacted soon, will contain specific provisions relating to area classification and the need to record the area classification information in an explosion protection document.

Hazardous area classification

In order to determine the location and extent of the hazardous areas it is necessary to establish the sources of release and the likely dispersion of the flammable material from the source in any direction to the point where its concentration is below the LEL. Factors such as the rate of release, concentration, volatility, ventilation, topography and density of the gas/air mixture will need to be taken into account. A contour line joining the points so established, both vertically and horizontally, is the boundary between adjacent zones.

BS EN 60079-10:1996 provides guidance on a systematic approach that can be used to identify the volumetric characteristics of the hazardous areas, taking due account of the factors listed above. The standard contains a number of example calculations and shows how the zones can be marked around possible sources of release (see Fig. 15.1 for an example).

Establishing the boundaries between the various zones in a hazardous area is a difficult task because of the assumptions and variables involved and it is best carried out by those concerned who have the appropriate knowledge, such as the process, production and chemical engineers and the safety officer. It will usually be found relatively easy to determine the Zone 0 and Zone 1 boundaries where the release of the material is controlled. These boundaries are usually within a few metres of the release source, but Zone 2 boundaries are more difficult because the emission is not controlled as it is due to an abnormal occurrence, such as a leaking pump gland or perhaps a burst pipe, and may occur in unmanned areas where immediate discovery is unlikely.

Generally, the Zone 0 and Zone 1 areas are situated within associated Zone 2s because an unintended release of the material will almost certainly extend beyond the Zone 0 and Zone 1 boundaries. It may also be found that zones overlap, necessitating, perhaps, the regrading of a Zone 2 to a Zone 1 area. Where the flammable material is heavier than air, any pits, trenches, cable ducts or cellars without forced ventilation and in Zone 2 areas should be designated Zone 1 as the flammable material will not readily disperse from within them. For lighter-than-air gases, a similar designation is appropriate for coffered ceilings and under roofs where the gas is likely to be trapped.

Individual companies, trade associations and professional bodies have issued guidance on area classifications for their particular sectors. For example, the Institute of Petroleum's and the Association for Petroleum and Explosives Administration's publication *Guidance for the design, construction, modification and maintenance of petrol filling stations* contains, at Chapter 3, comprehensive guidance on hazardous area classification at petrol filling stations. For petroleum and other flammable liquids in containers and bulk storage, reference should be made to HSE booklets:

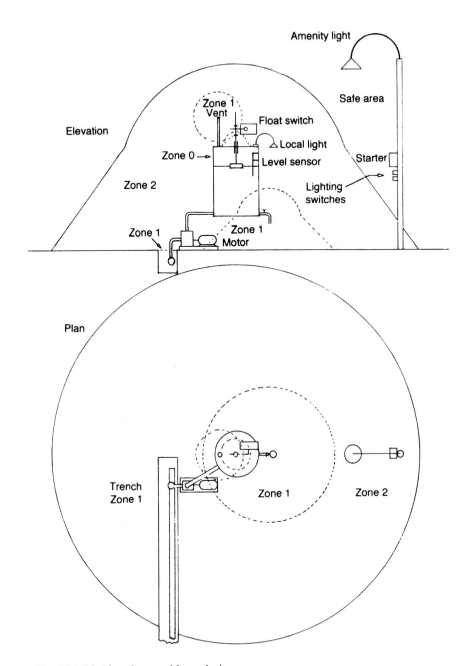

Fig. 15.1 Plotting the zonal boundaries.

■ HS(G)176 Storage of flammable liquids in tanks, which gives guidance on the design, construction, operation and maintenance of installations used for the storage of flammable liquids in fixed tanks operating at or near atmospheric pressure. It applies to new installations and to existing installations where reasonably practicable. It is relevant to industries such as chemical, petrochemical, paints, solvents and pharmaceutical. The guidance gives help in the assessment of the risks arising from the storage of flammable liquids, and it describes measures to control those risks such as containment (primary and secondary), separation, ventilation, sub-stitution, and control of ignition sources.

■ HS(G)51 The storage of flammable liquids in containers.

Gas groups

Engineers responsible for selecting electrical apparatus for installation in a hazardous area will need to know into which gas group the flammable atmosphere will fall. This is because, if either 'flameproof' or 'intrinsically safe' equipment is intended to be used, the grouping will affect the char-acteristics of the equipment. There are two main groups: Group I, which relates to electrical apparatus for use in mines susceptible to firedamp; and Group II, which relates to electrical apparatus for use in places where an explosive atmosphere exists, other than mines susceptible to firedamp. Group II gases and vapours are further subdivided into Groups IIA, IIB and IIC:

■ Group IIA representative gas: propane
■ Group IIB representative gas: ethylene
■ Group IIC representative gas: hydrogen

Temperature classification

Another consideration for electrical equipment is its temperature classifica-tion, or T rating. This indicates the maximum surface temperature of those surfaces of the equipment that can be exposed to the flammable atmosphere. So the T rating of electrical equipment should be below the minimum ignition temperature of the vapours or gases likely to be present in the flammable atmospheres. The T ratings in terms of maximum apparatus surface temperature are:

■ T1 – 450°C
■ T2 – 300°C
■ T3 – 200°C

- T4 – 135°C
- T5 – 100°C
- T6 – 85°C

Equipment categories

As a further complication, the ATEX Directive, enacted in the UK as The Equipment and Protective Systems Intended for Use in Potentially Explosive Atmospheres Regulations 1996 (see Chapter 7) introduced the concept of equipment categories, which will have to be adopted at the end of the transition phase in 2003. In very broad terms these categories are:

- Category M1 and M2 equipment – for mining use;
- Category 1, Category 2 and Category 3 equipment – for non-mining use.

Because of their limited application, the mining equipment categories will not be considered further.

The effect of the definitions, as far as electrical equipment is concerned, is that electrical equipment intended for use in Zone 0 areas would normally be classified as Category 1; equipment intended for use in Zone 1 areas would normally be classified as Category 2; and equipment intended for use in Zone 2 areas would normally be classified as Category 3.

EXPLOSION PROTECTED APPARATUS

Concepts

Having carried out an area classification exercise, the electrical equipment to be installed in the area must then be selected for its explosion protection properties. There is a range of options to choose from, with a variety of protection techniques employed. Each of the techniques, bar one, is described in detail in a European harmonised standard and is allocated a designation letter, mainly for ease of labelling. The following text summarises each of the techniques, and identifies its letter designation and appropriate construction standard.

Intrinsically safe – Type 'i' – EN 50020

The term 'intrinsically safe' applies to the apparatus and to the circuit in which is connected. It is normally associated with instrumentation circuits and some portable apparatus. This type of equipment may arc or spark under normal operating or fault conditions, but the design has to ensure that the energy level, even under fault conditions, is too low to produce an incendive spark that could ignite the flammable atmosphere. There are two categories, Type 'ia' and 'ib': ia is more stringent as it allows for two faults whereas ib allows for only one. In general, fixed intrinsically safe equipment installed inside a hazardous area is supplied through a device called a 'Zener Barrier' which serves to limit the energy delivered to the equipment in the hazardous area.

Pressurised or purged equipment – Type 'p' – EN 50016

This technique excludes the flammable gas from the apparatus interior by using slightly higher pressure inside than the ambient pressure outside. If the flammable gas is excluded, any internal arcs or sparks will not lead to ignition. In both cases the pressurising equipment has to be monitored and interlocked with the supply to disconnect it should the pressurising system fail. If the system has been shut down for any length of time, arrangements are necessary to purge it of any gas that may have entered, before the supply is restored.

Flameproof enclosures – Type 'd' – EN 50018

Flameproof equipment is totally enclosed. The flammable gas/air mixture may enter the enclosure and be ignited by any internal arcs or sparks that may occur. However, the enclosure has to be strong enough to withstand an internal explosion without damage and without allowing the internal explosion products to leave the enclosure at a temperature that could ignite the flammable atmosphere outside. To prevent such an ignition the flame paths through joints have to be sufficiently long to act as flame traps. Joints are generally flanged or labyrinthine, e.g. threaded. There are strict requirements laid down in the standards about the lengths of such joints and the maximum gaps that can exist between joint surfaces.

Increased safety – Type 'e' – EN 50019

Increased safety equipment is more akin to ordinary non-explosion-protected designs, but special precautions have to be taken to prevent any electrical or frictional sparking or hot spots exceeding the specified temperature limits. To prevent excessive heating the equipment is liberally rated and under fault conditions the associated control gear has to interrupt the circuit before dangerous overheating occurs.

Non-sparking apparatus – Type 'N' – EN 50021

Type 'N' apparatus is designed so as to be non-sparking in normal operation and such that a fault capable of causing ignition is unlikely to occur.

Oil immersed apparatus – Type 'o' – EN 50015

In this type of technique, the electrical apparatus or parts of it are immersed in oil in such a way that an explosive atmosphere which may be above the oil or outside the enclosure cannot be ignited.

Encapsulation – Type 'm' – EN 50028

In this type of protection, electrical parts which are capable of igniting a flammable atmosphere by either sparking or heating are enclosed in a compound.

Powder filled apparatus – Type 'q' – EN 50017

In this type of protection, an enclosure containing electrical parts which are capable of igniting a flammable atmosphere by either sparking or heating, is filled with a material in a finely granulated state. The design ensures that any arc produced by the electrical apparatus surrounded by the material cannot ignite the surrounding flammable atmosphere.

Special protection – Type 's'

Type 's' protection refers to any other protection technique not satisfying the requirements of the formal standards listed above, which a manufacturer can demonstrate to be safe.

Combinations

Some manufacturers produce equipment that combines different techniques into one package, usually in the interests of economy and ease of use. A common combination, for example, is Type 'd' and Type 'e' protection. The flameproof (Type 'd') enclosure is used to protect the sparking contents of the apparatus, and the lighter Type 'e' protection is used for the terminal box.

SELECTION OF EXPLOSION PROTECTED APPARATUS

Not all explosion protected electrical apparatus can be used in all zones of hazardous areas. This is because in some protection techniques the integrity of the technique is not sufficiently high to deal with the risk. The following limitations apply:

Apparatus for use in Zone 0:	Ex ia
Apparatus for use in Zone 1:	Ex d
	Ex p
	Ex q
	Ex o
	Ex e
	Ex ia and ib
	Ex m
	Ex s
Apparatus for use in Zone 2:	Ex N
	All those suitable for use in
	Zones 0 and 1.

Equipment approval and certification

Historically, there has been no duty on users of electrical equipment in flammable atmospheres to select equipment that has been approved or certified for use in such environments. The duty on employers, under the Health and Safety at Work Act, section 2, and the Electricity at Work Regulations, Regulation 6, has been to select equipment that is safe, so far as is reasonably practicable. However, guidance published by the HSE and others is that the best way of ensuring compliance is to use equipment that has been approved or certified against published standards by an appropriate test body. Similarly, suppliers of such equipment have been advised to ensure that their products are tested by an independent third party against the relevant standards before placing them on the market.

In the UK, the type testing and certification work has been carried out by the Electrical Equipment Certification Services (EECS), which is part of HSE, and by SIRA Test and Certification Ltd. The Equipment and Protective Systems Intended for Use in Potentially Explosive Atmospheres Regulations 1996 will change this, because they introduce a legal requirement for compliance with the Essential Health and Safety Requirements (EHSRs) and the application of the CE mark using the applicable conformity assessment procedures, which may involve the participation of a notified body. This is a complicated area but, in summary:

■ Category 1 electrical equipment and protective systems will have to be subject to EC type examination by a notified body for conformity to either a harmonised standard or the EHSRs or a combination of the two, plus either production quality assurance or product verification procedures.

- Category 2 electrical equipment will have to be subject to EC type examination by a notified body for conformity to either a harmonised standard or the EHSRs, plus either production quality assurance or conformity to type.
- Category 3 electrical equipment will have to be subject to internal control of production by the manufacturer.
- Alternatively to the above procedures, each production item may be subject to examination by a notified body for conformity to either a harmonised standard or the EHSRs or a combination of the two.

The overall aim of this is to ensure both that equipment is safe and that it is not subject to artificial trade barriers within Europe.

Labelling

Manufacturers of certified explosion-protected electrical equipment, fix labels to their products to indicate that it has been tested and approved for use in flammable atmospheres, and to provide details of the approval. The label identifies the manufacturer; the name or type of the apparatus; the year of construction; the construction standard against which the apparatus has been tested; the protection concept(s) used; the apparatus grouping (gas group) if applicable; the name or mark of the certifying body (notified body under ATEX); the T rating; and any other relevant information, such as the IP rating. The label mark starts with the Ex prefix, or EEx if a harmonised European standard has been used as the basis of the approval. As an example, the label 'EExd IIB T6' on a piece of apparatus indicates that it is of flameproof construction (Type 'd') for use in atmospheres of gas group IIB in which the gas has an ignition temperature greater than 85°C. There will also be the distinctive EC trading mark – Epsilon followed by 'x' within a hexagon symbol – as well as the CE mark if the apparatus has been approved in accordance with the provisions of the ATEX Directive.

THE ELECTRICAL INSTALLATION

Armed with the hazardous area classification plan, and knowledge of the applicable standards, the designer can design the electrical installation. In doing so he should avoid the hazard wherever possible by locating the electrical equipment and wiring outside the hazardous areas. This can be done, for example, using the following methods:

■ Luminaires may be situated in a safe area outside and lighting a flammable zone interior through lay lights in the roof or walls.

■ Motors can be located in a safe area, driving a machine within the flammable zone by means of a shaft extending through a sealed gland in an impermeable partition.

■ Motor control gear can often be located in a safe area remote from the motors in a flammable zone. For many processes where there is centralised control it is often an advantage to locate the starters in motor control centres, which can usually be situated in a safe area.

However, electrical apparatus and wiring that has to be in the flammable zones must be explosion protected to avoid the ignition hazard. Two examples are provided in the following sections.

Paint shop installation

Paint shops that use flammable or highly flammable paints must have explosion protection concepts incorporated in the design of the electrical installation. The paint shop installation shown in Fig. 15.2 has some electrical apparatus within the hazardous area, which would have to be explosion protected according to the zone designation. The non-ventilated pit would be Zone 1 and the pit lights would have to be Zone 1-compatible; EEx p type luminaires are shown in the diagram, although Eex d luminaires are commonly used. Above ground level there would be a mixture of Zone 1 and Zone 2 areas. In the immediate vicinity of where the flammable paint was being sprayed, there would be a Zone 1 volume, but a metre or so above the highest spraying level it would be Zone 2 assuming the paint solvent vapours are heavier than air. The motor would therefore, for example, have to be an EEx d type, or the purged EEx p type shown.

Both the motor and pit lights need safeguards against the entry of flammable gas when not in operation. Purge cocks are provided on the pit luminaires; these should be opened and then the compressor started to purge any gas from within them and the pipe ventilated motor. The cocks are then closed which raises the pressure in the pipeline and closes the pressure switch, which in turn makes the contactor switching circuit and enables the pit lights to be switched on. The flow of air through the outlet pipe of the pipe-ventilated motor closes the flow switch and energises a tilting device in the motor starter. After the prescribed interval the motor starter becomes operational and the purged motor can be started. The emergency stop buttons can be explosion protected or could be non-electrical (pneumatic or hydraulic) connected to a pressure switch in the safe area which in turn is connected to the starter 'stop' circuit.

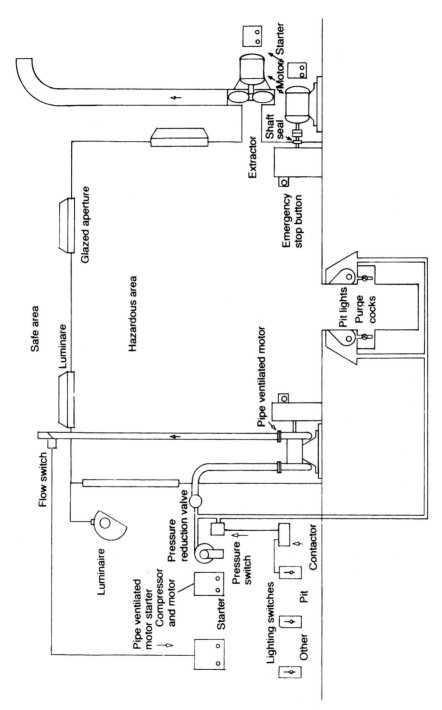

Fig. 15.2 A flammable paint spray shop for painting vehicles.

Tank of flammable liquid

The example in Fig. 15.1 shows the classified zones in the vicinity of a tank of flammable liquid which emits a heavier-than-air vapour. The space above the tank liquid is normally classed as Zone 0 because sufficient air may be drawn in through the vent and the float switch tube when the liquid level falls or the vapour condenses on a fall in the ambient temperature, to form an explosive mixture. The level sensor would normally be intrinsically safe type EEx ia.

Defining the Zone 1 and Zone 2 boundaries is not an exact science because there are many variables; it is more a matter of judgement. The extent of the Zone 1 areas around the vent and float switch tube depends on the rate of emission of the flammable vapour from these two orifices. A reasonable approach might be to work on the average release rate, assuming both a 100% vapour concentration and the plume being blown in one downward sloping direction by the wind, and then calculate how far it would disperse to reach the lower explosive limit. The average rate would be perhaps the mean between zero at low ambient temperatures and the maximum attained at high ambients with solar heating of the tank when it was being filled. It is usual to assume a spherical shape for the Zone 1 boundary about the vent orifice, so the distance found would be the sphere radius.

The float switch is on the edge of the Zone 1 boundary, but it is doubtful whether the expense of an EEx d switch is justified. A weatherproof, hermetically sealed, mercury-in-glass, type Ex s switch would be a suitably economic alternative.

The emission from the tap-operated dispenser is dependent on the surface area of the exposed liquid when filling containers and the area of the container orifice. Again, it would be reasonable to assume that the vapour plume is blown in one downward sloping direction and disperses until the LEL is reached. The container orifice to this position is the radius of the Zone 1 area. The slope of the zone boundary is a function of the vapour density. The heavier the vapour, the steeper the slope.

The pump motor is partly within Zone 1 so it would probably be advisable to use a weatherproof EEx e rather than an EEx N motor.

The extent of the Zone 2 area is dependent on the possible mechanical failures of the pump, float switch, tank or associated pipework. The worst situation is probably an overflow through the vent and float switch tubes, following a float switch failure and resulting in a fairly rapid spillage of the contents on to the ground. This would provide a large area of evaporation and the windblown plume would disperse across the ground surface for a considerable distance before attaining the LEL. Using this distance from the tank as the radius, the extent of Zone 2 can be established. The Zone 2 boundaries from a leaking pipe or damaged pump gland are likely to be less extensive and need not, therefore, be calculated in this case.

The pipe trench, although in the Zone 2 area, is classed as Zone 1 because vapour could be trapped in it for a long time.

The local light should be suitable for Zone 2 and could be a weatherproof type EEx N.

The electrical apparatus outside the Zone 2 boundary does not need to be explosion protected but would have to be weatherproof.

Wiring

To avoid ignitions from the wiring installation, the necessary precautions include the following:

- The conductors should be adequately insulated for the declared voltage and to prevent leakage or short circuits between conductors or to earth.
- The conductors should be further protected against mechanical damage.
- The conductors should be protected against excess current.
- MIMS cable systems should be protected by surge arresters to prevent insulation failure if transient voltage surges are likely.
- The cables should be liberally rated to suit the load cycle without getting hot enough to damage the insulation or ignite a flammable atmosphere.
- The installation should be suitable for the environmental conditions apart from the flammable atmosphere, e.g. a wet location or a corrosive or dusty atmosphere.
- Where it is necessary to exclude the flammable atmosphere, cables, conduits and trunking and in some cases ducting should be sealed to prevent gas being transmitted through the interstices of the cable or along conduits, trunking or ducts.
- The metalwork of conduits and trunking and the armouring and screening of cables should be bonded and earthed.
- Cables should, preferably, be connected only at apparatus terminals. If intermediate joints are made, they should be of the encapsulated type, or a terminal box sealed with compound should be used.
- Joints in steel conduit should be of the threaded type, tight fitting and painted to exclude moisture so as to avoid corrosion and an adverse effect on earth continuity.
- Joints between different metals, widely separated in the electrochemical series, should be avoided where the junction may attain a sufficient potential difference to cause corrosion. Where there is a risk, the joint should be made watertight by painting or other treatment.

- Terminations of explosion-protected apparatus should be such as not to invalidate the apparatus certificate.
- Flexible metallic conduit should have a non-metallic inner sheath or be suitably designed so that movement cannot cause abrasion of the cables.
- Flexible metallic conduit is not a suitable protective conductor. A flexible cable conductor should be used for this purpose to ensure earth continuity between the fixed conduit and apparatus, and the flexible conduit joints should be earth bonded at both ends.
- Avoid the use of light metal and light metal alloy components, such as aluminium and aluminium alloy conduit, and accessories where frictional contact with oxygen-rich items, such as rusty steel, might occur and cause sparking. Alternatively, prevent contact with a plastic sheath or other covering.
- Avoid compression glands on cables covered with insulation and sheathing which have a cold flow property where the gas has to be excluded.
- Overhead telecommunication or power lines should terminate outside the flammable zone, be fitted with surge arresters and run underground into the flammable zone to minimise danger from voltage surges caused by lightning.
- Do not use cable sheaths as neutral conductors.

PORTABLE AND TRANSPORTABLE APPARATUS

Portable and transportable apparatus should be excluded from the hazardous areas, as far as possible, because the conditions of use make it, and particularly its flexible cable, a greater ignition and accident risk than fixed apparatus. When it is used, it should be frequently inspected, tested and maintained to avoid trouble. The flexible cable should be of the screened type. The connection to the supply should be by means of an EEx plug and socket, interlocked with the supply switch, to ensure that the plug can only be inserted or withdrawn when the circuit is dead. It is also recommended that the circuit is protected by a residual current circuit breaker or circulating current earth monitoring protection or both. A sound approach is to ensure, via suitable management procedures, that such apparatus is only used in a hazardous area after the area has been proven by measurement to be gas-free – work would then usually be carried out under a permit to work issued by the process specialist who certifies that the area is gas-free.

GAS DETECTOR PROTECTION

Another means of protecting against the explosion hazard in flammable atmospheres is to employ gas detectors. The principle is that the gas detectors detect flammable gas/air mixtures at a low percentage, typically 10%, of the LEL and cause an alarm to sound. If the concentration reaches 25% or so of the LEL, the system automatically initiates a shutdown of the process. This type of technique is commonly used in process plants such as oil and gas refineries, LPG storage tank farms, chlorine doping rooms in water treatment plants, and on vehicles such as electric fork lift trucks operating in Zone 1 and 2 areas.

An example of gas detector protection is found in unattended cold stores and cooling plants which use ammonia as the refrigerant. There is practically no risk in attended plants because a few parts per million of ammonia in air is easily detectable by the pungent smell, so the attendant is aware that there is a leak and can take remedial action before the concentration becomes dangerous. The lower explosive limit is comparatively high at 16%; this concentration is only likely under abnormal conditions, such as a blown compressor head gasket, and is intolerable for the eyes and respiratory system.

Adequate protection may be obtained by the use of gas detectors, suitably located near the ceiling, at positions where flammable concentrations may be expected from gas leaks. The detectors are connected to a controller which can be set to trigger an alarm, trip the circuit breaker controlling the supply to the electrical equipment in the machine room and, if required, switch on an explosion-protected extractor fan to disperse the gas at a concentration of about 2%, or to actuate the alarm and fan at this low concentration and trip the circuit breaker at a higher value of say 7%. The protection will operate well below the LEL of 16%. The integrity of the protection system can be ensured by providing it with a standby supply for automatic connection in the event of its mains supply failing. The relevant standard is BS 4434 Requirements for refrigeration safety: Part 1: 1969 General, and the principles are depicted in Figs 15.3 and 15.4.

Figure 15.4 is a single line circuit diagram. The gas detectors GD1 and GD2, their controllers and the alarm are powered from the supply or from an inverter which is battery energised so an alarm would be given should a gas leak occur when the plant is shut down or the supply fails. Under normal conditions, the extractor fan is controlled by the non-explosion protected two-way switches in the safe area and compressor room, but in the event of a leak gas detector GD1 would operate. This would shut down the compressor room installation and automatically disengage these switches and their associated non-pro-

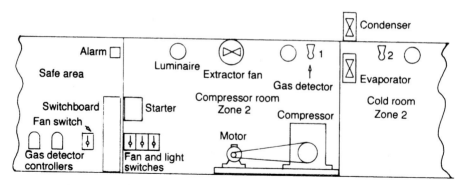

Fig. 15.3 Gas detector protection for ammonia refrigeration plants.

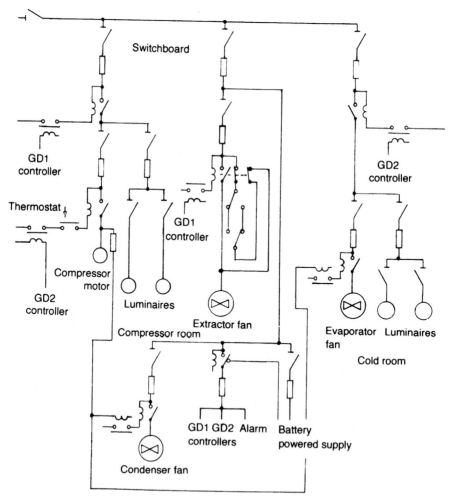

Fig. 15.4 Ammonia refrigeration plant – single line circuit diagram.

tected wiring and connect the extractor fan to the supply via wiring of a type specified in the British Standard.

A leak in the cold room would be detected by the sensor GD2 which would energise the alarm and shut down the cold room installation. Figure 15.4 also shows that an interruption of the compressor motor supply would shut down the condenser and evaporator fans. This, however, is an economic rather than a safety feature.

Gas detectors can also be usefully employed in other unattended Zone 2 plant areas where it is desirable to trigger an alarm in the event of a large gas leak which, if not detected and remedied, might spread beyond the zonal boundaries and reach a source of ignition in an adjoining safe area.

AVOIDANCE OF DUST EXPLOSIONS

There is a considerable range of materials, from powdered aluminium to flour, that will explode when mixed with air in a dust cloud. An explosion will occur on ignition if the cloud is within flammable limits. Whereas gas/air mixtures will disperse with time, dust clouds tend not to disperse. Moreover, layers of dust can accumulate on surfaces, which can present a permanent fire risk, and when disturbed they can give rise to a cloud explosion risk. To prevent ignition, the principle is to separate the dust cloud from the source of ignition by making the electrical apparatus dust-protected to IP5X or dust-tight to IP6X. To achieve this the apparatus has to be totally enclosed, with sealed joints to limit or prevent the ingress of dust. As an alternative, type EEx p pressurised or purged apparatus may be used. The differential between the inside and outside pressures prevents the ingress of dust.

This pressurising method helps to prevent the bearing trouble that sometimes affects totally enclosed motors with sealed joints. However, these machines are subject to pressure cycling as they tend to breathe outwards when operating and warm, and inwards when shut down and cold. Some of this breathing may be through the bearing housings and may introduce dust from the polluted atmosphere which may contaminate and damage the bearings. A bearing failure could result in overheating and/or sparking and cause an ignition. Small amounts of dust penetrating through the joints, however, are not usually hazardous because the resultant dust cloud is below the LEL. The motors should be dismantled periodically and the dust removed.

Temperature

Totally enclosed equipment is often designed for a greater temperature rise than enclosed ventilated equipment. It should be selected so that the final temperature attained, i.e. ambient plus temperature rise, is less than the auto-ignition temperature of the flammable dust. Some allowance should be made for the increase in temperature rise due to the heat insulating effects of the dust coating on the apparatus which is likely to occur even with good housekeeping.

Totally enclosed fan-cooled motors are cheaper than totally enclosed machines of the same output and are, to a certain extent, self-cleaning as the air-flow over the body shell from the external fan prevents the build-up of the dust.

BS 6467 : 1988 Electrical apparatus with protection by enclosure for use in the presence of combustible dusts

BS 6467 has two parts. Part 1 is a specification for the apparatus and recognises two categories: IP5X dust-protected, which allows some dust penetration but the quantity is insufficient to cause an ignition, and 1P6X dust-tight, where no dust penetration is permitted.

Part 1

To avoid the risk from sparking, disconnecters, plugs and sockets, and the doors of distribution boards and switchfuses have to be interlocked with an isolator to ensure off load operation for fuse changing. Alternatively, a warning notice can be affixed. A similar notice should be used on luminaires to ensure no sparking risk when they are opened for relamping or other work.

The apparatus should be labelled with the IP category, maximum surface temperature and minimum temperature for use or storage if this is lower than $-20°C$. The reason for this is that low temperatures can embrittle some materials and lead to a cracking risk. The maximum surface temperature allows for a 5 mm layer of dust.

To avoid the risk of electrostatic ignitions, Class II apparatus should not be used and Class I equipment metalwork should be effectively earthed. Light metals and their alloys should be avoided where there is a risk of frictional sparking.

Apparatus exteriors should be designed to minimise dust accumulations and to facilitate cleaning. The maintenance instructions should include information on preserving the integrity of joints and gaskets to exclude dust, and on cleaning methods to remove the dust without creating a static charge.

Part 2

Part 2 of BS 6467:1988 Guide to selection, installation and maintenance, divides hazardous areas into two zones. Zone Z is where combustible dust is, or may be, present as a cloud during normal processing, handling or cleaning operations in a sufficient quantity to be capable of producing an explosive concentration of combustible or ignitable dust in a mixture with air. There may also be a dust layer. Zone Y is any area not classed as Zone Z in which accumulations or layers of combustible or ignitable dust may be present under abnormal conditions and give rise to ignitable mixtures of dust and air.

Note that, under the Protection of Workers Potentially at Risk from Explosive Atmospheres Directive, there is an attempt to align the zone definitions for gases and dust clouds according to the risk of the hazard being present. Thus, there are three new zone definitions relating to dust:

■ *Zone 20:* an area in which a combustible dust cloud in air is present continuously, or for long periods, or frequently.
■ *Zone 21:* an area in which a combustible dust cloud in air is likely to occur in normal operation occasionally.
■ *Zone 22:* an area in which a combustible dust cloud in air is not likely to occur in normal operation but if it does it will persist for a short time only.

The relationship to the more well-known Zones 0, 1 and 2 is clear. However, the new definitions do not relate to layers or accumulations of dust, which must be avoided or, where they cannot be avoided, the surface temperature of equipment must be reduced to a safe level.

The Equipment and Protective Systems Intended for Use in Potentially Explosive Atmospheres Regulations 1996 also introduce a requirement to label equipment as being suitable for dust-laden atmospheres.

Returning to the Zone Z and Zone Y definitions, in a properly designed plant where precautions have been taken to minimise the escape of dust and where there is good housekeeping, the Zone Z area should only occur in the immediate vicinity of the source of release and usually within a domed cylinder volume of 1 m radius from the source of release, the height of the cylinder being from the source of release to the floor. Zone Y, however, caters for abnormal escapes due to some failure of the apparatus such as a sprung joint, so the extent of Zone Y is taken as within 1.5 m of the possible source of release in vessels or equipment and 3 m vertically above it. IP6X protection is advisable in Zone Z areas and IP5X in Zone Y.

If ferromagnetic dust is present, precautions may be needed to avoid or minimise external electromagnetic fields generated by apparatus to prevent dust accumulating. Aluminium conductors within apparatus should be kept away from casing joints to avoid the possibility of severe arcing in the event of a fault burning through the joint and igniting the dust outside.

There should be a safety margin between the ignition temperature of the dust layer or cloud and the maximum temperature of the apparatus. For a 5 mm layer, the recommended maximum temperature of the apparatus is 75°C less than the dust ignition temperature and for a dust cloud the maximum should not exceed two-thirds of the ignition temperature. Most dusts have relatively high ignition temperatures so there should be little difficulty in complying with the requirement.

The wiring installation should be in accordance with BS 7671:2001. Conduit should be of the solid drawn or seam welded type with screwed joints to prevent dust penetration. The joints should be protected against corrosion to preserve earth continuity.

Record drawings of the zonal areas should be updated to reflect plant alterations which affect the zonal boundaries.

RADIO FREQUENCY INDUCTION HAZARD

The risk of RF induction is small and likely to occur only in a few locations. It should tend to diminish as more apparatus is made to comply with EMC (electromagnetic compatibility) standards. However, installations in the vicinity of radio, TV and radar transmitters and repeaters should be checked in accordance with BS 6656.

THE LIGHTNING HAZARD

Again, lightning is a comparatively rare source of ignition. Outdoor installations are more vulnerable than indoor ones. The risks can be minimised by means of lightning conductors installed to BS 6651.

THE ELECTROSTATIC HAZARD

Liquids, solids, vapours, gases and dust clouds can readily become electrostatically charged to a high voltage but, with the exception of large capacitors, generally store comparatively small quantities of charge. The charge may be positive or negative and will discharge into anything of opposite polarity which it touches or, if the potential difference created is high enough, when it is sufficiently close for a flashover to occur. The resultant spark may well be hot enough, i.e. contain sufficient energy, to ignite a flammable atmosphere as most gases require only fractions of a millijoule and most dust clouds only between 10 and 100 mJ, according to the material, for this to occur.

Creating an electrostatic charge

There are a number of ways of creating static charges but most of the problems arise from two of them: rapid mechanical separation of bodies in contact with each other, i.e. friction; or induction, where the charge is due to the presence of an electric field. Typical examples of the first are the finishing of paper and textiles by calendering rolls, or the separation of an insulating material web from the rollers of a coating machine, or the flow of materials in pipe lines. Examples of the second are the charging of an operator wielding an electrostatic paint or powder spray gun.

Detection

The presence of static can be detected by instruments that measure the voltage or field strength, or, more crudely, by a gas discharge lamp which will glow if held in the field, or by the use of a wide band radio receiver and listening for the crackling sounds produced by the field. In severe cases the sparking can be heard and seen.

Measuring is best done when the flammable hazard is not present, but if this is impracticable then the tester should use certified, explosion-protected instruments and carry out the work so as to avoid creating an incendive spark. Only qualified persons with an adequate knowledge of the hazards to avoid and an expert knowledge of the difficulties that arise in measuring should do this work. Such an expert should be able to measure with sufficient accuracy and do the calculations to determine whether or not the static hazard is severe enough to cause an ignition and merit expenditure on remedial action. The difficulty is that the appraisal can only take into account the plant conditions at the time. Subsequent changes can alter the static hazard, so where an ignition could cause expensive damage or could be a substantial danger to the operators it will probably be prudent to provide some precautions anyway.

Precautions (BS 5958 Code of Practice for control of undesirable static electricity)

Whenever possible, the hazard should be avoided by preventing the build-up of electrostatic charges. This can be done in the case of conducting material by earthing. The resistance to earth should be not more than 1 megohm, except for locations where the capacitance is low when a higher resistant value is permissible, e.g. 100 megohms where the capacitance of the body is not more than 100 pF (picofarads).

It is more difficult to remove the charges from insulating materials. In some cases this may be done by modifying the material to improve its conductivity. For example, flammable liquid fuels can be doped with additive and carbon black can be added to rubber during manufacture to reduce the resistivity of the material. Where this is undesirable, or uneconomic, it may be possible to carry out the process in a humid atmosphere where the water vapour acts as a leakage path to earth. This method is not suitable, however for water-repellent materials.

Limitation of energy

Limitation of energy is a method used in electrostatic spraying where a high impedance is inserted in the HV circuit to reduce the potential earth fault current and thus the spark discharge energy to a sufficiently low value to avoid ignition.

Ionisation

Another precautionary method is to ionise the air at the surface of the material to provide a conducting path. It is often used on coating machines. A row of needle-pointed electrodes is arranged near to the surface of the web. Alternate electrodes are energised at HV and the others earthed to ionise the air and discharge the static. In another system the ionised air is produced by a radioactive sealed source and blown on to the surface of the material under an earthed electrode. These devices are generically known as 'static eliminators' and may themselves need to be built using explosion protection techniques to prevent them acting as the source of ignition.

Ventilation

In difficult cases, ventilation should be used to weaken and disperse the flammable atmosphere so that it is below the LEL, thus preventing ignition by an incendive spark.

Inert atmosphere

Ignitions can also be prevented by the use of an inert atmosphere. Nitrogen and carbon dioxide are generally used. This method can be utilised where the process machines are designed to employ it, but otherwise it may be impracticable. It has been successfully used in oil tankers to inert the cargo spaces during tank washing.

Earthing in large spaces

Flammable dust clouds can acquire a dangerous spatial charge which can spark over to an earthed electrode such as a measuring conductor, sampling tube or poking rod protruding into the container. The danger is a function of the volume, and ignitions are only likely in large spaces such as large storage silos and hoppers. The dust clouds occur during filling when probes should not be allowed in the cavity to avoid ignitions or, alternatively, the space should be divided by earthed partitions, rods or wires to prevent the build-up of HV spatial charges. Earthed partitions are also advisable in large tanks containing flammable liquids to prevent HV charging of the contents.

Earthing of mobile containers

Mobile containers on non-conducting rubber-tyred wheels, such as aircraft, road tankers and industrial trolley tanks and vats, can acquire static charges and should be discharged by means of an earthing lead before their flammable contents are handled. The earthing lead should be connected by a clamp with an insulating handle and the lead should next be earthed at a position remote from the flammable material. This practice serves the dual purpose of avoiding ignitions from an incendive spark and an unpleasant shock to the operator.

Operator training

Operators can initiate ignitions by unwittingly causing incendive sparks in flammable atmospheres. For example, if an operator who is well earthed by standing on a conducting floor or being in contact with earthed metalwork seeks to touch a charged surface, he can enable an incendive spark to jump the gap to his finger immediately before contact. Alternatively, an operator wearing insulating footwear or standing on an insulating floor can become charged and an incendive spark is possible when he goes to touch an earthed object. In circumstances where the latter hazard is significant, operators should wear clothing made of natural, rather than man-made, fibres to reduce the generation of charge, and should wear anti-static footwear to allow charge to leak away to earth before it reaches a hazardous level. In this case, measures should be taken to ensure that the floor has semiconducting properties – the charge will not relax to earth if the floor is non-conducting.

Operators in hazardous areas where static can occur need suitable training if ignitions are to be avoided.

Speed reduction

The generation of static charge is often related to the speed of the process, such as the speed at which a highly flammable liquid such as toluene is dispensed from a pipe into a vessel. A reduction of speed may be a sufficient precaution, in some cases, to reduce the static charge to a harmless level. This may be achieved, for example, by using larger bore pipes and thus decreasing the flow rate for the same output. In the case of webs and belts, a reduction in the linear speed or an increase in the diameter of rollers and pulleys will reduce the static charge.

MAINTENANCE

To ensure safety in flammable atmosphere locations, maintenance staff should be suitably trained and competent to carry out the necessary inspection, testing and repair work. This means, for example, that an electrician has to be trained and qualified in electrical technology and also the techniques peculiar to hazardous atmosphere apparatus and wiring installations. There are now industry schemes in place to provide electricians and technicians with the competencies they need for this type of work.

One such scheme is known as COMPEX. This is a national training and assessment scheme for electrical tradesmen who work in potentially explosive atmospheres. The scheme was jointly developed by the Engineering Equipment and Materials Users Association (EEMUA), the training arm of the Electrical Contracting Industry (JTL) and the National Electrotechnical Training Organisation (NET, formerly EIEITO). On successful completion of practical and written assessments of each unit, candidates are awarded a certificate of core competency which carries national recognition among EEMUA members.

The scheme comprises six units which are delivered via a network of approved centres and may be studied individually or in pairs covering installation and maintenance. The six training units which make up the scheme are:

■ Unit EX01 Preparation and installation of Ex 'd' N 'e' and 'p' systems
■ Unit EX02 Maintenance and inspection of Ex 'd' N 'e' and 'p' systems
■ Unit EX03 Preparation and installation of Ex 'ia' and 'ib' systems
■ Unit EX04 Maintenance and inspection of Ex 'ia' and 'ib' systems

■ Unit EX05 Preparation and installation of equipment protected by enclosure for use in the presence of combustible dusts
■ Unit EX06 Maintenance and inspection of equipment protected by enclosure for use in the presence of combustible dusts

Since 1996 three additional units to the Level 3 S/NVQ 'Installation and Commissioning Electrical Systems and Equipment' have been accredited by National Council for Vocational Qualifications (NCVQ) and Scottish Qualifications Authority (SQA):

■ Install and verify electrical installations for use in potentially explosive atmospheres.
■ Commission electrical installations for use in potentially explosive atmospheres.
■ Diagnose and rectify causes of electrical fault to restore effective performance on equipment installed in potentially explosive atmospheres.

Persons who have successfully completed any four of six COMPEX assessments and in addition can provide appropriate evidence of industrial experience can claim the NCVQ/SQA unit.

Chapter 16
Electric Arc Welding

INTRODUCTION

The arcs and sparks produced by arc welding and cutting tend to make the process appear more dangerous than it actually is. There are, in fact, comparatively few reported electrical accidents involving welding activities. Whereas the operators undoubtedly experience occasional electric shocks, such injuries are not a serious hazard at the low voltages employed for most types of welding. In confined, conductive and/or wet locations, however, the typical electrode open circuit voltage of 80 V a.c. can be lethal and there have been occasional electrocutions. Operators are also prone to minor burn injuries from contact with hot metal and to 'arc eye' from ultraviolet radiation. The accidents are almost wholly confined to manual welding where an operator holds an electrode holder or torch.

The most common process is a.c. manual metal arc welding where flux-coated stick electrodes are used and the power source is a transformer with inductive control of the welding current between about 50 and 600 A. The secondary windings of the transformer are isolated from the casing and hence from earth, as shown in Fig. 16.1. The standard open circuit voltage is 80 V, falling on load to some 20 to 40 V depending on the load current.

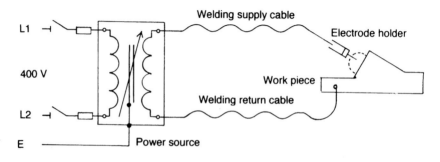

Fig. 16.1 Manual metal arc welding.

The non-electrical safety precautions comprise protective clothing for the operator to prevent burn injuries from arc sputter and the ultraviolet and infrared radiation produced by the arc. A visor is used to protect the face, with an aperture fitted with a filter glass through which the operator observes the area. It is usually necessary to provide extract ventilation to remove fumes, and opaque or filter screens, arranged around the operator, to protect other people in the vicinity from ultraviolet radiation (see Fig. 16.2).

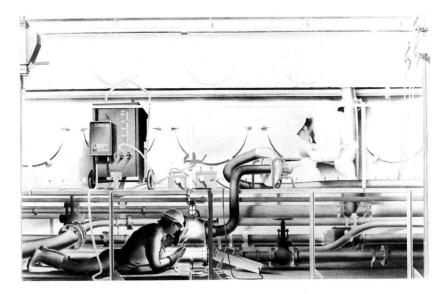

Fig. 16.2 Welding under a walkway in a ship's engine room. The safety precautions are: (1) The power source is fitted with a low voltage device. (2) The handlamp and heated quiver are fred from step-down transformers and operate at safety extra-low voltage. (3) There is an insulated box for the electrode holder when not in use. (4) The fumes are extracted by the centrifugal fan. (5) The operator is wearing insulated clothing and rubber boots and is separated from the steel floor by lying on dry wooden planks. (6) There is a second person present.

ELECTRICAL SAFETY PRECAUTIONS

Earthing

From an historical perspective, in order to avoid indirect contact electric shock, the now-superseded 1966 edition of BS 638 Specification for arc welding plant, equipment and accessories, recommended that the power source metalwork and the work piece should be earthed. This was a safe-guard against an interwinding fault in a transformer source causing the welding circuit to become live at the primary voltage. If this type of fault were to occur, the earthing arrangements would lead to the operation of the overcurrent protective device (such as a fuse or circuit breaker) on the supply to the welding transformer.

In 1991, BS 7418 Code of practice for isolation of the welding circuit in arc welding plant, was published. It advocated that the British practice of earthing the work piece be changed to align with the European practice of not earthing the welding circuit. There was an important proviso, however, that the power source transformer must comply with BS 638: Parts 9 or 10: 1990, or be as safe as a BS 3535 isolating transformer, to reduce to a negligible level the risk of an interwinding fault. This meant that the secondary windings should be isolated from the case of the transformer and that there should be reinforced or double insulation to prevent galvanic breakthrough from the primary windings to the secondary windings. This arrangement is now fully reflected in BS EN 60974-1 : 1998 Arc welding equipment – Welding power sources, and is the basis of current practice. BS 7418 has been withdrawn.

Before adopting the practice of not earthing the work piece, a user of a power source which predates BS 638: Parts 9 or 10: 1990 should determine, preferably by asking the manufacturer or supplier, whether or not the transformer is as safe as a BS 3535 transformer. Some of the existing older transformers and transformer rectifiers will not comply, and in these cases either the welding transformer should be replaced with one that does meet the more modern specification, or the practice of earthing the work piece should be continued.

Users of the older transformers should note that it is not advisable to save the cost of the welding earth cable by earthing the welding circuit at the transformer as this provides a possible parallel return path for the welding current and can lead to damage from stray currents. Figure 16.3 illustrates a special combined earthing and welding return clamp, which ensures that the operator does not forget the earth connection to the work piece. When it is clamped on to the work piece, it connects it to earth and to the power source.

Welding circuits supplied from engine-driven generators do not need not to be earthed as there is no risk from mains voltage.

In those cases where the welding circuit is earthed and where the electricity supply is PME, there is a risk of damage to the low voltage service cable supplying the premises as, in the event of stray welding currents circulating in the metalwork in the premises, there will be a parallel circuit through the CNE conductor of the service cable to the nearby PME earthing points. This circuit will carry a share of the stray currents, which may damage it. To avoid this possibility, precautionary measures should be discussed with the supply company and may include using non-earthed welding circuits to present-day

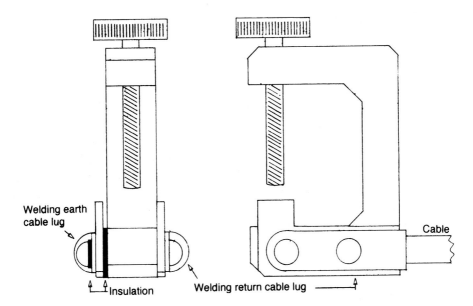

Fig. 16.3 Welding clamp.

standards, a residual current circuit breaker in the supply circuit, or an SNE supply to the premises. Of course, although the welding circuits may not be deliberately earthed, they are often fortuitously earthed because the work pieces are in contact with earthed metalwork in which the stray currents circulate. Damaging stray currents are more likely to occur where there is more than one operator, so to mitigate potential damage and a minor shock hazard, extraneous metalwork near the operator should be bonded to the work piece and nearby work pieces should be bonded together.

Large power sources

Large power sources are usually transformers and are used in shipyards and fabrication workshops, for example, where a number of operators are working at the same time. Sometimes the operators are supplied via one or more distribution boxes and individual regulators, as shown in Fig. 16.4. In this diagram the welding circuit protection has been omitted for simplicity but, although it may not be observed in practice, the distribution box should have circuit breakers with overcurrent protection or fuses to comply with Electricity at Work Regulations, Regulation 11, and to prevent a potentially damaging stray current in the event of a phase-to-earth fault on the welding lead between the distribution box and any one of the regulators. The protection is not needed for single operator sets or three-phase power sources

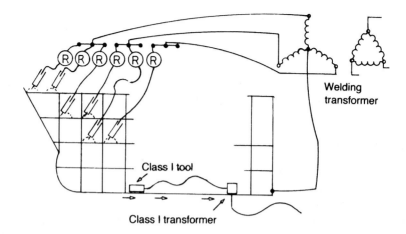

Welding
transformer

Class I tool

Class I transformer

Fig. 16.4 Parallel current path through protective conductor of Class 1 tool cord.

feeding only one operator/phase where the primary overcurrent protection should suffice, or when the secondary windings at the transformer are isolated from earth. The practice of using large transformers to supply multi-operators is, however, being superseded in favour of single operator sets where the regulator is incorporated in the unit and where the primary overcurrent protection is adequate.

Regulators and distribution boxes

Where a single power source is used to supply a number of welding circuits, as shown in Fig. 16.4, distribution boxes and regulators are employed to provide a regulated supply for each operator. This equipment is usually of heavy Class I construction and the metalwork is earthed to avoid indirect electric shocks.

In the event of an earth fault in a distribution box or regulator, its metalwork would become live. Where the metalwork and work piece are earthed, there would be no shock risk but the fault current would probably damage the protective conductor. If the work piece was not earthed, however, a potential, at the welding voltage, would exist between the accessory metalwork and work piece, but there would be no fault current unless they were in contact. As the regulators and distribution boxes are invariably close to the work piece, the shock risk can be avoided by the provision of an equipotential conductor between the equipment and work piece in place of the protective conductor which is often run with the welding return conductor and earthed at the power source. In the event of an earth fault this short equipotential conductor would probably be damaged by the fault current, but such damage would be evident and the cable could be replaced at no great cost. This would be less hazardous than damaging the protective conductor circuit and should be considered for use with both earthed and unearthed welding circuits. The problem does not arise when regulators and distribution boxes are of Class II construction, thus avoiding the need for a protective or equipotential conductor.

Electrode holders

Electrode holders should be insulated to prevent contact with live parts as far as possible. There are two types of electrode holder specified in BS EN 60974-11 : 1996 – Types A and B. Type A is less versatile than type B but somewhat safer as no live parts are accessible to the standard test finger and when the electrode is fitted, its non-coated end does not protrude from the electrode holder head. To avoid direct electric shocks, operators should not change electrodes with bare hands unless the power source is switched off.

A box made of incombustible insulating material should be provided for the operator to rest the electrode holder when not in use to avoid the possibility of extraneous metal being made live by contact with a live electrode stub and/or causing stray currents. When the task is completed and/or the operator is not present, the power source should always be switched off.

Confined conductive spaces

The 80 V a.c. open circuit voltage can be dangerous in hot or wet and/or confined and conductive spaces, i.e. in any situation where the impedance of the operator's body to earth is low. In these circumstances the prospective shock current could easily exceed the threshold value of 0.5 to 2 mA and may be sufficient for electrocution, so special precautions are required to minimise the shock hazard. The possible electrical safety measures are to use a d.c. power source with a smoothed output, i.e. not more than 10% ripple, at the lowest practicable welding voltage (a rectifier power source should be so designed and constructed that a rectifier fault would not cause a dangerous voltage or ripple increase); or an a.c. power source with a low voltage control device, as illustrated in Fig. 16.5. This control device interrupts the welding supply on open circuit so that only a signal voltage of 20 to 30 V is present at the electrode. To weld, the operator touches the work piece with the electrode, and the control responds by restoring the welding supply, so enabling the arc to be struck. Withdrawing the electrode breaks the arc, causing the control to interrupt the welding supply and to restore the signal voltage. The device thus ensures that the maximum voltage at the electrode holder held by the operator cannot exceed the maximum arc voltage of about 40 V on load

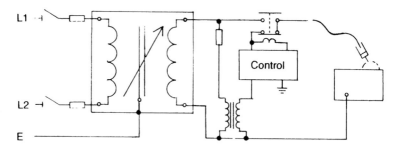

Fig. 16.5 Low voltage control device circuit.

or the signal voltage of 20 to 30 V on open circuit. In both cases, it is preferable for the power source to be located outside the confined, conductive space, otherwise the power source supply circuit should be protected by a sensitive RCD to protect the operator against both direct and indirect electric shock at mains voltage.

Any electrical accessories such as a hand lamp or heated quiver should operate at SELV. A heated quiver is a device used for holding electrodes at a temperature between 100 and 150°C to help protect them from absorbing moisture into the coating.

Non-electrical supplementary precautions are advisable, such as waterproof and/or insulating clothing in wet locations to keep the skin dry and out of contact with conducting surfaces to maintain a high body impedance. In hot locations, where the skin could be wet with perspiration, operator cooling fans may help, together with insulating sheet material to separate the operator from conducting surfaces. Such fans should operate on SELV or be out of the operator's reach. In confined conductive locations insulating material may also be used to separate the operator from conducting surfaces.

Higher voltage hazards

There are potential electric shock hazards at higher voltages from three-phase multi-operator welding and from the use of multiple power sources (see Figs 16.6 and 16.7). In these cases the operators should be separated so that no operator can touch two electrode holders at different voltages at the same time.

Stray currents

For older transformer power sources, if the welding circuit is earthed at the power source instead of at the work piece, a parallel return path may result which will carry a proportion of the welding current dependent on the

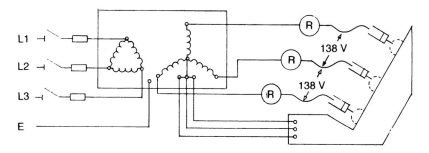

Fig. 16.6 Three-phase open circuit.

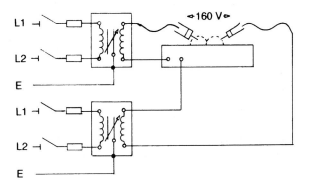

Fig. 16.7 Possible open circuit electrode voltages from two power sources.

relative impedances of the two paths. If the return clamp is not connected, the parallel path, through the extraneous metalwork and/or protective conductors, will carry all the welding current of several hundred amperes and may well cause considerable damage (see Fig. 16.8). A loose or dirty connection may also cause the parallel path to carry a dangerous proportion of the welding current.

Welding sets with earth-referenced secondary windings at the transformer should be regarded as obsolete and taken out of use to prevent the stray currents causing damage.

Other welding processes

MIG and MAG

In MIG (metal inert gas) and MAG (metal active gas) systems the welding electrode is an uncoated wire, fed from a gun held by the operator. Oxidation of the weld is prevented by shielding it with a gas dispensed via the gun. The

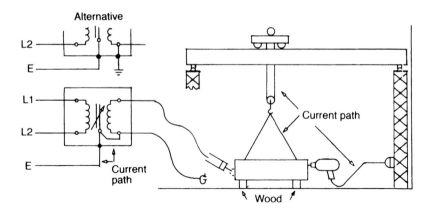

Fig. 16.8 Possible stray welding currents.

wire feed unit may be separately mounted or incorporated in the power source. It is motor driven and controlled by a switch on the gun. Where the work piece is fortuitously or deliberately earthed, the problem is to prevent the electrode wire in the wire feed unit from making contact with anything earthed, thereby shorting out the arc. This usually entails an unearthed control circuit and the use of Class II or Class III apparatus. Figure 16.9 shows a typical example of apparatus complying with BS 638. It is advisable to provide the wire feed unit with an insulating cover to IP4 of BS EN 60529 : 1992 to prevent the live parts being accidentally touched by any earthed metalwork in the vicinity and to protect personnel from a direct shock at the open circuit voltage.

TIG

TIG (tungsten inert gas) welding employs a tungsten electrode which does not come into contact with the work piece. The arc is struck and maintained by ionising the gas between the electrode tip and work piece with a high voltage which is sometimes at a high frequency. To avoid the HV shock hazard from accidental contact with the electrode, the energy content of the HV circuit should be controlled so that the maximum possible shock current is limited to about 0.5 mA.

Plasma

Plasma arc welding and cutting can be of either the transferred or non-transferred arc type. The arc is struck between a non-consumed centre electrode and an annular tube surrounding it through which the gas is emitted. In the transferred arc process, which is applicable only to con-

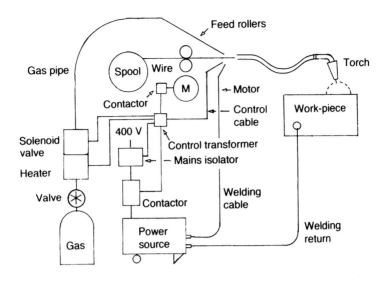

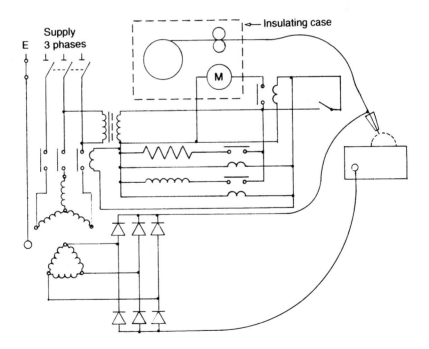

Fig. 16.9 MIG/MAG welding.

ducting work pieces, the arc is transferred after striking from the tube to the work piece. The torch is normally water-cooled. The cooling pipes, welding circuit conductor, arc initiation circuit and control circuit cables and a pipe for the gas are all usually contained within a common flexible pipe which terminates at the torch. Higher voltages are used particularly for the machine process but, for the manual type where there is a potential shock hazard if the plasma tip is touchable, voltages are restricted to 200 V for cutting and 220 V for gouging. When the arc is extinguished the open circuit voltage must not exceed an a.c. peak of 68 V and 48 V r.m.s. which means the provision of a low voltage control device as shown in Fig. 16.5. Provided that the equipment is properly maintained and water leaks are avoided, it is relatively safe when used by a trained operator.

Heated quivers

The relevant standard for heated quivers is BS 638: Part 5: 1988 which requires the maximum supply voltage to be 110 V and a maximum voltage to earth of 80 V. Class I quivers have to have an earthing conductor and a fuse to protect the device. Heated quivers are crude ovens used to keep electrodes warm and prevent them absorbing moisture from the air. A quiver consists of a steel box containing a metal sheathed heating element (see Fig. 16.10). The more primitive types have a spiked connector and an insulated magnet, except for the contact face, for connecting the element to the welding cable and work piece, respectively. When the operator moves his work station he pulls the magnet off the work piece and receives an 80 V shock if he is in

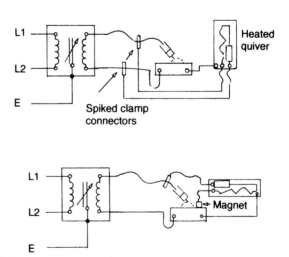

Fig. 16.10 Heated quiver connections.

contact with the uninsulated part of the magnet and work piece or other bonded metalwork at the same time. Some of the devices are fitted with two spiked connectors, one for fixing to the welding and the other to the return cable, which overcomes the shock problem but, when the connectors are removed from the cables, the cable sheath punctures are revealed so this method of connection is not recommended. The preferred method is shown in Fig. 16.11. If it is necessary to use the device in a confined conductive location it should be operated at SELV (see Fig. 16.12).

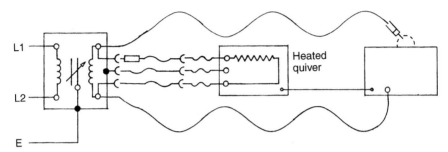

Fig. 16.11 Recommended method of connecting a heated quiver.

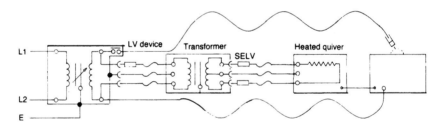

Fig. 16.12 Heated quiver circuit in confined conductive locations.

Heated quivers, like regulators, are invariably used close to the work piece and can, therefore, be bonded to the work piece by a short equipotential conductor in place of the protective conductor for indirect shock protection in the event of an earth fault.

SUPERVISION, TRAINING AND MAINTENANCE

The hazards of electric are welding can be avoided provided that there is adequate safety supervision to ensure that the appropriate precautions are taken, particularly when work has to be done in confined/conducting, wet or hot locations; that operators have been instructed in the safety measures; and that the apparatus is periodically checked by the operator and supervisor to ensure that it is maintained in a safe condition. This includes seeing that all connections are of ample size to carry the welding current and are clean and tight. Care is necessary in the use of magnetic clamps to ensure that the contact surfaces are clean and flat and that there is a sufficient area in contact to avoid overheating.

Chapter 17
Tests and Testing

INTRODUCTION

Electrical apparatus, equipment and installations are tested to prove that they are safe and fit for purpose and that they comply with relevant standards, specifications and codes of practice.

The need for testing, and its extent and type, varies through the life cycle of a particular item. During development, prototypes of apparatus and equipment need detailed and extensive testing to prove the design, both in terms of functional safety and inherent electrical safety; this testing is usually done in the laboratory. Once the design has been proved and the item is put into production, rather less elaborate testing is needed during manufacture to ensure that each product is safe and meets the specified requirements. For more complex equipment, testing at various stages of assembly on the production line is common to detect faulty components and to minimise the failures of the completed product when it is tested before despatch. Further testing is then needed during the in-service life of the item as part of routine preventive maintenance to ensure its continuing safety integrity. Additional testing is needed after the item has been subject to any repair, refurbishment or modification.

Some test facilities, such as those provided for high power short circuit testing, are specialised and expensive and are usually provided by testing authorities to whom manufacturers submit their prototype products for testing in preference to providing the facility themselves. Where products have to be certified, e.g. explosion-protected apparatus, it is usually necessary to have the testing done by an approved laboratory, or notified body, which may itself issue compliance certificates or the certificates may be provided by a certifying authority for whom the laboratory works. Other test facilities used for routine testing of products are less complex and the precautions to be taken against electrical injury are well-defined, not least in the European Standard EN 50191 : 2000 Erection and operation of electrical test equipment.

Electrical installations are usually tested on completion and before commissioning, but large installations may be tested and commissioned in sections as each part is completed. Subsequently, they need periodic retesting as part of the maintenance procedure to detect incipient faults and to prevent the development of fault combinations that may lead to danger.

This chapter describes, firstly, the tests routinely carried out on low voltage installations; and, secondly, the testing of apparatus and equipment, and the precautions against electrical injury that should be taken in test facilities.

INSTALLATION TESTING

As explained in Chapter 10, the recognised standard for extra-low and low voltage installations is BS 7671:2001. Part 7 of the standard provides recommendations for the inspection and testing of installations, with amplification published in the IEE's Guidance Note No. 3 – Inspection and Testing. The information in these publications has much in common with what follows in the next few pages. The standard stipulates that the results of inspections and tests should be recorded in an inspection and test certificate which, together with the circuit diagram and equipment manufacturer's instructions, should form part of the operational manual and always be available to the installation user and whoever carries out the periodic inspections. Compliance with the recommendations in Part 7 and Guidance Note 3 is one way of demonstrating compliance with the Electricity at Work Regulations, Regulation 4(2) on maintenance.

Inspection

When an electrical installation is completed and before it is energised, a visual inspection should be carried out to ensure that the installation complies with the specification and with BS 7671 and that the connected apparatus and wiring are undamaged. Strictly speaking, inspection does not fall under the heading of 'testing' but it is an essential prerequisite. Both BS 7671, Regulation 712-01-03, and IEE Guidance Note No. 3, provide a checklist that should be used to ensure that nothing is missed. Some items merit particular attention, such as:

■ In the connections of conductors, strands are not cut back to accommodate one or more conductors in an inadequately sized terminal, there are no loose terminal screws and wiring insulation is not cut back too far.
■ The conductors are of adequate size not only for current rating but also for voltage drop.

- The conductors are labelled where necessary to facilitate subsequent maintenance.
- The cables, conduit and trunking which pass through walls or floors designated as fire barriers are sealed externally and internally.
- The wiring is not located where it may be adversely affected thermally, e.g. in the vicinity of radiators or covered by thermal insulation, unless its current rating and insulation is adequate and suitable.
- The wiring and apparatus are suitable for the environmental conditions, e.g. corrosive, polluting, wet or dusty atmospheres, which may occur in parts of the premises and which may not have been appreciated when the specification was prepared.
- Isolating and control devices, e.g. the positions of emergency stop push buttons, for potentially dangerous machinery, are correctly located.
- The labelling of switches, fuses and circuits is suitable.
- The earthing terminals of Class 1 equipment are connected to the protective conductors.
- The installation is earthed and the main earthing conductor is the correct size.
- The main equipotential bonding conductors are present and of the correct size.
- Cable glands are properly made off, wire armouring and braiding is securely held by the gland and dust boots are correctly fitted.
- Supplementary equipotential bonding is provided where necessary.
- Fuses are of the correct type and rating and excess current trips correctly set.
- Drainage holes are provided at low points in conduit systems.
- Flexible metallic conduit is not used as the protective conductor.
- The wiring diagram or schedule is correct and the operational manual adequate and complete.

Whereas the foregoing checks are for pre-commissioning inspections, routine visual inspections of the installation should be an integral part of the in-service maintenance regime. In addition to the checks listed above, any signs of overheating or other distress of the components making up the installations should be used as an indication that something may be wrong and that a detailed investigation of the problem is warranted.

When the person carrying out the inspection is satisfied with the results, the tests listed in BS 7671, section 713, should be carried out, following the recommended sequence to ensure that it is safe to energise the circuits for the final three live tests. The main tests are described in the following sections.

Continuity tests

Protective conductors

It is not necessary to measure the continuity of phase and neutral conductors in radial circuits as an open circuit is self-evident, but the continuity of protective conductors, including the main earthing conductor and the main and supplementary equipotential bonding conductors, has to be verified. For conductor sizes not exceeding 35 mm^2, their inductance, which is small, may be ignored and the testing can be done with a low resistance d.c. ohmmeter. It should be capable of reading to two decimal places and the results should be recorded for subsequent use in calculating the earth loop impedance and for comparison purposes when carrying out periodic inspection and testing. Where, however, conductor sizes exceed 35 mm^2, an a.c. low impedance ohmmeter, of similar resolution, should be employed. It should operate on SELV but be capable of providing a test current of 1.5 times the design current, but it need not exceed 25A.

Radial circuits are tested by connecting the phase and protective conductors together at the distribution board and then applying the test between the phase and earth terminals at each outlet point. A higher than expected reading should be investigated and will probably be due to a defective connection. Where the disconnection times are found to exceed those of Table 41A of BS 7671, it will be necessary to measure the protective conductor resistance on its own.

To measure the continuity of the earthing conductor and the main and supplementary equipotential bonding conductors, their ends remote from the consumer's earthing terminal should be temporarily disconnected to avoid parallel paths, and a test applied between these remote ends and the main earthing terminal.

IEE Guidance Note No. 3 points out that where there is an insulated protective conductor, run in metal conduit, trunking, MIMS and armoured cables, it should be tested during erection and before the item supplied is in place so as to avoid a parallel path through the exposed conductive parts. Where, however, the exposed conductive parts are used as the protective conductor and are secured to the building steelwork or other extraneous conductive parts, there may well be parallel paths. In these cases, the impedance or resistance of the exposed conductive parts is likely to compare favourably with that of the phase conductor because of their greater cross-sectional area, provided their terminations are properly connected and the design is correct, so any parallel paths through extraneous metalwork are a bonus. However, a test should be done and if the reading is less than the designed amount, it can be regarded as satisfactory. If there is any doubt

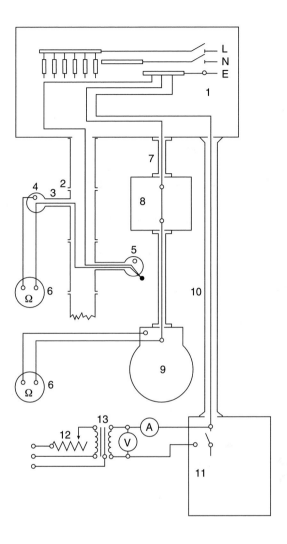

Key
1. Metal-cased distribution board
2. Cable trunking
3. Conduit
4. Conduit box for luminaire
5. Conduit box for lighting switch
6. Ohmmeter
7. MICC cable with terminating glands
8. Motor starter
9. Motor
10. Armoured cable with terminating clamps and glands
11. Boiler
12. Rheostat
13. Step-down transformer

Fig. 17.1 Measuring the earth continuity of exposed conductive parts.

about the integrity of the connections, the a.c. high current ohmmeter should be used for the test, or a test equipment of the type shown in Fig. 17.1 should be employed and the resistance calculated.

Continuity of ring final circuit conductors

The first test to be done consists of joining the phase and neutral conductors of opposite ends of the ring at the distribution board, as shown in Fig. 17.2a, and measuring the resistance between the other two to prove continuity of the phase and neutral conductors. A high resistance reading indicates an open circuit or wrong connection. These other two are then connected and the ohmmeter connected between the phase and neutral conductors at each socket outlet in the ring (see Fig. 17.2b). Each reading should be about the same provided that there are no interconnecting loops. The method is repeated connecting the phase and protective conductors, as shown in Fig. 17.3. The reading taken at the socket outlet at the ring mid-point is the loop resistance to be used in calculating the earth loop impedance of the ring circuit.

Earth electrode resistance

The method of measurement of earth electrode resistance is indicated in Fig. 17.4 and is appropriate if an electricity supply is available. Otherwise a self-powered portable instrument should be used and the manufacturer's instructions followed. The current and voltage electrodes, which are usually 13 mm diameter steel rods, are driven into the ground by means of a vibrating hammer to a depth of about 1 m; if the contact resistance between the ground and electrode is too high, it should be driven deeper and/or watered with brine. The current electrode has to be outside the resistance area of the electrode under test. A distance of 30 m apart should ensure this in most cases. The voltage electrode is first located midway between the other two at position 1 in Fig. 17.4, and the current and voltage read. Readings are then obtained with the voltage electrode relocated at positions 2 and 3. For accurate results, the voltmeter should be a high resistance type so that its operating current is a negligible proportion of the current between the earth and current electrodes. Provided the readings are about the same, the earth electrode resistance is the mean voltage divided by the current.

Figures 17.5 and 17.6 show the voltage readings plotted against distances, where it will be seen that the maximum voltage drop occurs near to the electrodes. Where the resistance areas do not overlap, the curves have a horizontal section and the three readings are sufficient to establish the earth electrode resistance. Where the conductivity of the ground varies between the

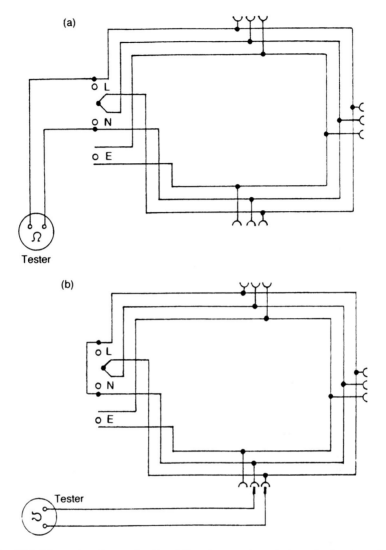

Fig. 17.2 Measuring the continuity of the phase and neutral conductors in a ring circuit.

electrodes and/or their resistance values are significantly different, the shape of the curve and the resistance areas of the electrodes will vary and perhaps overlap. In this case the separation distance between the earth and current electrodes should be increased and the test repeated. The horizontal section of the curve may not be midway between them so additional voltage readings should be taken if necessary nearer to each electrode to establish the location of the horizontal section of the curve. If the readings fluctuate, it is due to the

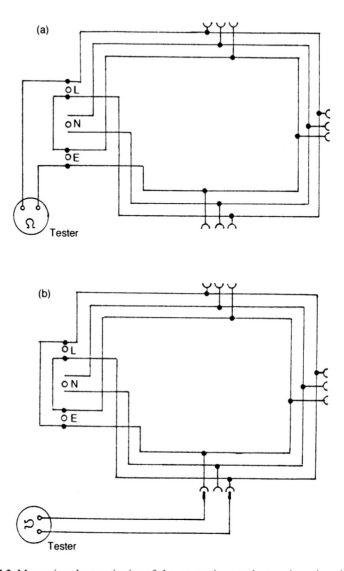

Fig. 17.3 Measuring the continuity of the protective conductors in a ring circuit.

presence of stray currents of the same frequency in the ground. In this case, the results will be unreliable and the test should be carried out at a different frequency. This is best done by using a portable purpose-made earth electrode tester.

In IEE Guidance Note No. 3, an alternative test method is given for RCD-protected TT installations. This consists of connecting an earth loop impedance tester between the earth electrode and the phase conductor of the installation, at the intake. Before doing this test, all equipotential bonding

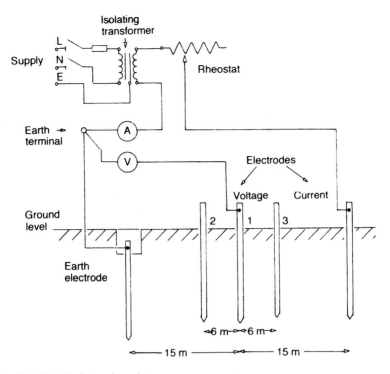

Fig. 17.4 Earth electrode resistance measurement.

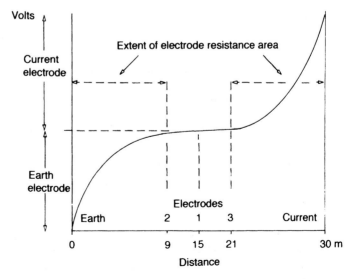

Fig. 17.5 Voltage/distance curve where the electrode resistance areas do not overlap.

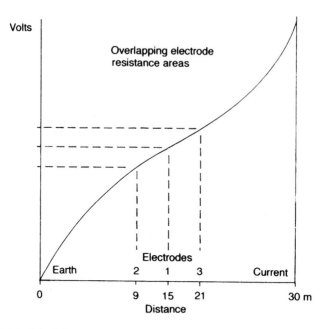

Fig. 17.6 Voltage/distance curve where the electrode resistance areas overlap.

conductors are disconnected temporarily to ensure that all the test current passes through the electrode. The earth loop impedance so obtained is treated as the electrode resistance for calculation purposes.

Insulation tests

BS 7671 specifies that d.c. test voltages should be used for insulation resistance tests, and that the minimum acceptable insulation resistance values should be in accordance with Table 17.1. The instruments used should be capable of providing an output of not less than 1 mA. For installations designed to operate at voltages up to 500 V, e.g. those connected to the 400/230 V public supply and the 110 V systems supplying portable apparatus, a 500 V d.c. insulation tester is adequate and suitable for the prescribed test. A 1000 V tester should be used to test systems over 500 V up to 1000 V. Large installations should be tested in sections, each of not more than 50 outlets. An outlet, for example, consists of a lighting point, a switch or a socket outlet.

As the object is to test the installation and not the connected apparatus, which should have been tested by the maker, it should be prepared by removing lamps and, for discharge lighting, unplugging or disconnecting one pole of capacitors or starting devices, also one pole of bell transformers,

Table 17.1 IEE advocated d.c. test volages and minimum acceptable insulation resistance values

Circuit (Volts)	Test (Volts)	Insulation resistance (megohms)
Extra-low, i.e. not exceeding 50 V a.c. or 120 V d.c.	250	0.25
Over extra-low and up to 500 V	500	0.50
LV over 500 V but not exceeding 1000 V 1000 V a.c. or 1500 V d.c.	1000	1.00

indicator lamps or anything else connected in parallel with the supply. Disconnect connected apparatus, such as a cooker, by opening the double pole switch in the control unit and, for fixed apparatus controlled by a single pole switch, open it and disconnect the neutral. Voltage-sensitive equipment such as dimmers and any other susceptible electronic gear should be protected from damage by connecting the terminals together.

Insulation resistance between current-carrying conductors and earth

The first test is between the current-carrying conductors and earth and is done by connecting the phase(s) and neutral conductors together at the supply point and measuring the insulation between them and the consumer's earthing terminal. Both 'on' positions in two-way switched circuits should be tested (see Fig. 17.7). This test is not, of course, applicable to TN-C installations.

Insulation resistance between current-carrying conductors

The second test is to prove the insulation between the current-carrying conductors and is carried out by testing between each phase/neutral and the other conductors connected together at the supply point (see Fig. 17.8).

Suspect insulation

In most cases, the insulation readings will far exceed the minimum tabulated values, but where a reading is less than 2 megohms, each circuit should be separately tested and if any is less than 2 megohms it should be checked for a latent defect.

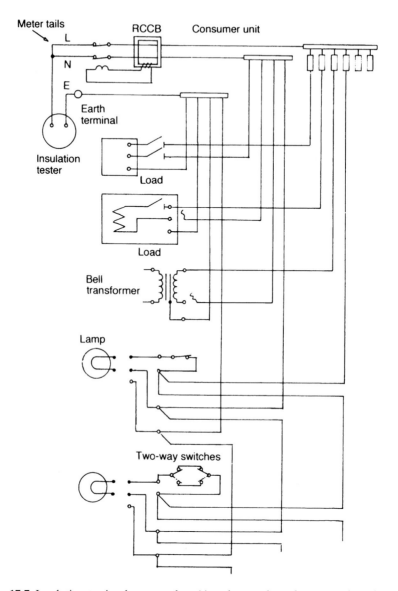

Fig. 17.7 Insulation testing between phase(s) and neutral conductors and earth.

Site-applied insulation

Where equipment is fabricated or repaired on site and the work involves the insulation of live conductors, the insulation standard should be comparable with that of similar factory-built apparatus and be subjected to the same HV testing. In the case of LV switchgear and controlgear to BS EN 60439 : 1994, the test voltage, between live parts of different polarity and between live parts

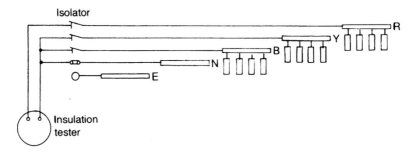

Fig. 17.8 Insulation test between each phrase or neutral conductor and the other conductors connected together.

and earth, varies according to the system voltage and is 2000 V for a 230 V system and 2500 V for a 400 V system applied for 1 minute. Lower voltage control circuits are tested at voltages shown in Table 17.2.

Table 17.2 Test voltages for low voltage control circuits	
Circuit volts U_i	**Test voltage**
Up to 12	250
12 to 60	500
Over 60	2 U_i + 1000, minimum 1500

There is also an impulse voltage test, applied for three cycles, which is dependent on the main circuit voltage. For 400/230 V systems it is as shown in Table 17.3.

Table 17.3 Impulse test voltages for 400/230 V systems		
Type of supply	**Test volts (kV) Location of equipment**	
	At intake	Elsewhere
Overhead line	6	4
Underground cable	4	2.5

Where a live conductor has been wrapped with insulation which is touchable, the test is done by wrapping metal foil around the applied insulation and testing between it and the conductor. If there is no relevant British Standard the test voltage is 3750 V a.c. at 50 Hz, applied for one minute.

Gaps in barriers or enclosures have to be shielded to prevent the entry of fingers (IP2X), and the top entry of wires (IP4X) making contact with the live parts.

SELV circuits

If the installation includes SELV circuits, for battery charging for example, it is essential to prove that there is no possibility of the SELV circuits being energised at a higher potential because the live parts of such circuits may be handled. This possibility exists where the power source is a transformer and/or where the SELV cables are not separated from cables operating at a higher voltage. The IEE recommended tests comprise a 500 V d.c. insulation test applied for one minute between the input and output circuits, the insulation resistance to be not less than 5 megohms, and finally, an insulation test at 500 V d.c. between the SELV circuit and the protective conductor of the supply input circuit.

Walls and floor in non-conducting location

If the installation includes any special facilities such as a test bench where it is necessary to establish and maintain a non-conducting location, the walls and floors must be made of non-conducting material. Their insulation value has to be proved by measuring the leakage current between at least three points on each relevant surface and earth. So a test bench against a wall would require a test at not less than three points on the adjacent wall and three on the floor. If there is an extraneous conductive part in the location, such as a metal window frame or service pipe, one of the test points should be not less than 1 m or more than 1.2 m from it. Check that the extraneous conductive part is separated from any exposed conductive part so that they cannot be touched simultaneously. If they can, the section of the extraneous part, within reach, has to be insulated to withstand a 2 kV a.c. test with a leakage current not exceeding 1 mA.

Although the standard gives no guidance on the test probes to be used, which should simulate the contact of hands and/or feet with the test surfaces, IEE Guidance Note No. 3 describes two types of electrode, one square and the other triangular. The square one has 250 mm sides and is used with a square of damp paper or cloth between it and the test surface. The other is a tripod with three circular electrodes, spaced at 180 mm, each having a contact area of 900 mm. The electrodes are in a flexible material such as conducting rubber. They are applied to the test surface which is either moistened or covered with a damp cloth. In both cases, the electrodes are pressed against the test surface with a force of 250 N (76 kg). Where the test surfaces are uneven, a more resilient type of electrode with a contact area of 250 × 250 mm, such as that illustrated in Fig. 17.9, may prove to be more suitable.

One test lead should be connected to the probe and the other to the nearest protective conductor. The test voltage should be at least 2 kV applied for one minute and the leakage current must not exceed 1 mA (see Fig. 17.9). This is followed by a 500 V d.c. insulation resistance test applied in the same way and at the same positions. The reading should be not less than 0.5 megohms for voltages up to 500 V or 1 megohms for voltages over 500 V.

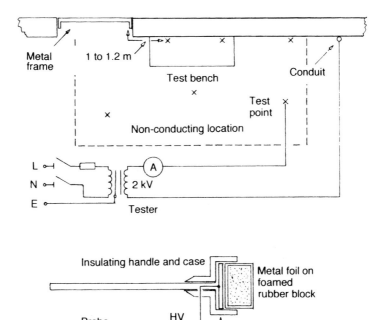

Fig. 17.9 HV testing of a non-conducting location.

Polarity test

There have been a few cases where the phase and neutral of single-phase TN-S supplies have been interchanged at the intake, so for a new installation it is worth checking that this has not happened. If the polarities are correct, a voltmeter connected between the conductor labelled 'phase' and anything earthed will display the mains voltage.

So far as the installation is concerned, the object of the polarity test is to prove that the fuses and single pole control devices are in the phase conductors only and that socket outlets and Edison screw lampholders are correctly wired. The continuity tests, already carried out, have proved that the socket outlet ring circuit connections are correct so no retest is needed. For the remaining radial circuits, IEE Guidance Note No. 3 recommends using the same number one test method employed for continuity testing, i.e. connect the phase conductor to the protective conductor at the distribution board and then test between the phase and earth terminals at each outlet. The two tests can, therefore, be combined, but for outlet substitute point in order to check, for example, that a switch between the distribution board and outlet is in the phase and not in the neutral conductor.

Loop impedance testing

Loop impedance tests are carried out to determine the loop impedance between the power source(s) and the point in the installation where the test is done. The device employed measures the current which passes through a resistor and displays the result in ohms. It is used to determine the loop impedance between phases, phase to neutral or any phase to earth. Some instruments incorporate a transformer to enable the neutral/earth loop impedance to be measured. From Ohm's law these readings can then be expressed in prospective short circuit fault currents.

The object of measuring the phase/phase or phase/neutral loop is to check that the short circuit rating of the protective devices is adequate. It should not be necessary to carry out this test on a new installation where the short circuit rating of the selected apparatus should be suitable for the calculated values of short circuit current, but the measurement is a useful check. The measurement, however, can only indicate the value when the reading is taken. Subsequent alterations in the supply network and connected apparatus can vary the value, so there should be a substantial safety margin between this ascertained value and the short circuit rating of the protection to allow for reductions in the loop impedance and a consequent increase in the prospective short circuit current.

The earth loop impedance should be measured so as to enable the prospective fault currents to be calculated. These currents have to be sufficient to operate the protective fuses or circuit breakers within the specified time in the event of an earth fault. Again, the readings obtained indicate the current position, which may alter subsequently. In most cases load growth tends to increase the prospective fault current, but in areas subject to dereliction the load may decrease and the consequent network alterations may result in a fall in the prospective fault current.

The readings are taken at the end of a circuit remote from the protective device. In a socket outlet ring main they are taken at the socket outlet nearest the mid-point of the ring and at the end of non-fused spurs. The readings are then converted into prospective fault currents by Ohm's law, i.e. $I = E/Z$, where E is the phase-to-earth voltage. These currents are then used to ascertain the corresponding protective device's operating time from the time/current curves of the protective device.

The measurements should be taken between a phase and earth, not neutral and earth, and at a current exceeding 10 A. Instruments which are readily available operate at 20 to 25 A where the supply system is TN-S or TN-C-S and the earth loop impedance is low, but usually at a much lower current on TT systems where high earth loop impedances are common. In these cases the operating current may be under 10 A and if the installation has a protective conductor which is mainly steel conduit, the reading will not be accurate. A second test is made, therefore, of the earth loop impedance external to the installation by measuring between one of the incoming phase terminals and the main earthing terminal with the main equipotential bonding conductors disconnected. The effective value of the earth loop impedance is then taken as twice the first reading less the second. This method must be used also where an instrument of the rapidly reversed d.c. type is employed and the installation is in steel conduit.

In circuits protected by RCDs and where the current exceeds the RCD tripping current, carrying out the test may trip the RCD. Some test instruments are designed to prevent this happening by, for example, first passing d.c. current to saturate the magnetic core of the core balance transformer in the RCD. Where such an instrument is not available, or it fails to prevent the RCD tripping, the measurements can be taken after defeating the RCD's tripping mechanism or bridging the phase terminals. It may not be practicable to do this safely on some RCDs built into consumer units. In this case the earth loop impedance is measured at the incoming mains location as described above, and the total earth loop impedance for any point in the installation is taken as the summation of this value and the resistance of the phase/earth loop as measured in the continuity tests.

Bridging the phase terminals of an RCD will prevent its inductance being included when the earth loop impedance is measured, and the calculated resultant fault current will be slightly higher than the true value. This, however, is immaterial as in the event of an earth fault the RCD will trip anyway within the required time. It is, of course, axiomatic that any bridging conductors are removed once the tests have been completed.

Figure 17.10 shows a domestic installation connected to a TT system of supply and an earth loop impedance tester connected to measure the earth loop impedance external to the installation. It will be seen that the loop consists of the phase conductor from the incoming terminals of the consumer unit via the metering and supply company's fuse to the HV/LV

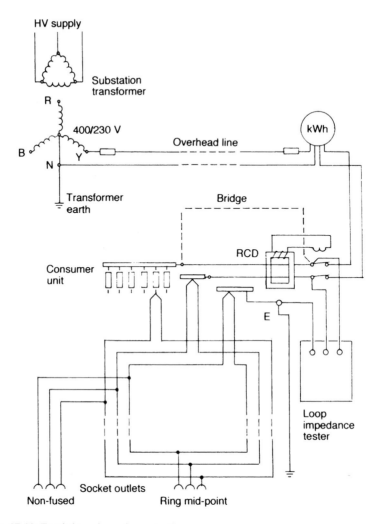

Fig. 17.10 Earth loop impedance testing.

distribution transformer's secondary winding, then via the transformer neutral and earth electrode, through the ground to the consumer's earth terminal. If the RCD trip can be defeated or bridged, the earth loop impedance at the socket outlets shown can be measured by connecting the instrument at these points. Otherwise, the calculation method described above will have to be used. When carrying out phase/earth loop impedance tests, the protective conductors, the metalwork of Class I apparatus and bonded extraneous metalwork are energised and there will be a potential difference between them and the ground. For high earth loop impedances, this would be a substantial proportion of mains voltage, so to avoid any shock hazard no one should touch the protective conductors or metalwork during the test.

RCD testing

BS 7671 stipulates that RCDs should be tested with their loads disconnected by pressing the test button and also by supplying an earth leakage current and checking the tripping time. Figure 17.11 shows a small installation where the RCD is a circuit breaker and serves also as an isolating switch. The second test can be safely carried out by the following procedure:

(1)   open the RCD;
(2)   remove the fuses;
(3)   connect the tester between the incoming side of a fuse carrier and the earth busbar;
(4)   close the RCD;
(5)   operate the tester.

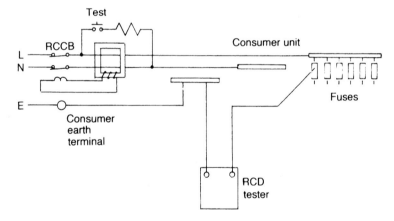

Fig. 17.11 Checking the tripping time of a residual current circuit breaker.

The specified requirement is that the rated tripping current shall cause the circuit breaker to operate within 0.2 s or at the delay time declared by the device maker. Where protection is against direct contact the device should operate within 40 ms for a residual current of 150 mA. On larger installations, where there is a switchboard feeding a number of distribution fuseboards, for example, earth leakage protection may be provided on the incoming and/or outgoing circuit breakers as well as on the final sub-circuits. All the devices should be tested and any designed discrimination between them verified.

The operation of the tester will cause the protective conductors, Class 1 apparatus metalwork and bonded extraneous conductive parts to become live with respect to the ground at a voltage dependent on the earth loop impedance, so during the tests no one should touch such metalwork.

Excess current protection testing

BS 7671does not require the testing of excess current protection, presumably because circuit breakers should have been tested by the maker. However, where excess current and the operating time of the trip are set on site, it is advisable to verify the performance using primary or secondary current injection apparatus. Such testing is essential on high voltage systems, where excess current protection is usually provided by relays fed from current transformers. Incidentally, such apparatus can conveniently be used to test earth leakage protection where the leakage trip current and time are outside the limits of the normal RCD testers.

ELECTRICAL APPARATUS AND EQUIPMENT TESTING

There are very many circumstances in which electrical apparatus and equipment need to be tested; some examples are:

■ Portable, transportable and fixed equipment, as well as installations, being tested as part of a routine preventive maintenance regime. The legal duties relating to maintenance of equipment were set out in Chapter 6 and the general topic of maintenance is covered in Chapter 18.
■ During fault finding on electrical installations and equipment.
■ Appliances and equipment being tested on a production line.
■ Motors being tested after having been rewound.
■ Computer monitors and televisions after having been repaired.

The tests will be a combination of functional tests and electrical safety tests, the latter being aimed at ensuring that the equipment will not be electrically dangerous when energised. This will principally involve testing the insulation resistance characteristics of the device and ensuring the integrity of the earthing arrangements on Class I appliances, although many other types of tests are carried out depending on the application (e.g. measurement of magnetising losses, thermal effects, in-rush currents, harmonics, electrical and acoustic noise levels, radiation levels and so on).

The risks of electrical injury during testing can be inherently high because conductors energised at dangerous voltages are often exposed to touch. It is essential, therefore, that measures are taken to ensure that the persons who are undertaking these tests are not exposed to unacceptable levels of risk, and the following text describes the main techniques for achieving this.

The relevant legal requirements are the Electricity at Work Regulations, mainly Regulations 4(3) and 14. It should always be borne in mind that the legal duty is that the work should always be done dead unless it is unreasonable for the conductors to be dead, the level of risk is acceptable, and suitable precautions are taken against injury. A very common misconception is that testing does not constitute a work activity and therefore does not come within the remit of the Regulations. This is, of course, not the case – testing is a work activity so the full panoply of the law applies.

In those situations where testing must be done with the conductors live, the most important requirement is that the tester is protected against direct contact with those conductors energised at dangerous potentials. There are two main ways in which this can be achieved:

(1) the use of interlocked jigs and enclosures to keep people away from the live conductors during testing;
(2) the use of techniques such as limitation of energy, electrical separation and earth leakage protection, coupled with precautionary measures such as the use of barriers and warning indicators.

Jigs and enclosures

Testing during the course of manufacture on an assembly line or before and after the repair of components can often be done using interlocked jigs and enclosures. The general principle is that the item being tested is placed inside the enclosure, perhaps after test probes and instrumentation have been attached to the item, and the enclosure door is then closed. Only once the door is closed can energy be applied to the item under test and the action of opening the door would lead to the supply being disconnected. One such type of interlocked jig, and the associated interlocking device, was described in Chapter 13.

Enclosures used for this purpose should generally be constructed with an ingress protection rating of IP3X, unless access to the live parts is otherwise safeguarded by, for example, the use of electrosensitive protective equipment. Moreover, the interlocking system should be designed so as not to fail to danger in the event of a single fault occurring, meaning that the safety-related parts of the control system should satisfy the category 3 principles set out in BS EN 954-1 (see Chapter 13).

Washing machine test

Figure 17.12 is an example of the way in which enclosures can be used to achieve safety during product testing. The example used is that of a washing machine test. The product, a partly assembled washing machine, arrives at the roller conveyor where the operator anchors the flexible cable to the product frame and connects it. The product then goes through the tunnel which has a slot in the roof for the flexible cable. When the product is out of reach it operates normally open limit switch no. 1 and the contactor closes. At about the same time the normally open limit switch no. 4 is closed by the rotating cam on the spring drum and the signal lamps are lit to warn

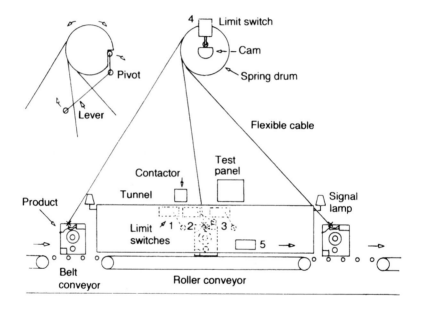

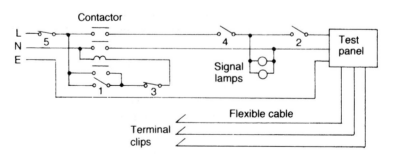

Fig. 17.12 Automatic electrical testing of products on an assembly line.

operators that energy has been applied. Then limit switch no. 1 opens but the contactor stays closed as its coil circuit remains energised through the auxiliary contact. Normally open limit switch no. 2 closes and energises the test panel which supplies the product through the flexible cable and automatically carries out the test programme. The test panel can be designed to print out the results and/or initiate an alarm if the product is faulty. Next, limit switch no. 2 opens, normally closed limit switch no. 3 opens and the contactor opens. The signal lamps go out and limit switch no. 4 also opens.

The inset shows a ratchet on the spring drum as a further refinement. When the product is leaving the tunnel the ratchet is engaged and prevents the spring drum's rotation and the closing of limit switch no. 4. When the operator anchors the cable on the product frame at the tunnel entrance, the cable catches the end of the lever and releases the ratchet which can only reengage when the product reaches the tunnel exit. In this type of system, the terminal clips should be insulated. With this arrangement the operator is well safeguarded against the possibility of touching any live parts.

Live testing

Live testing of equipment, apparatus and installations is a common cause of electrical injury and frequently requires careful consideration and planning of the precautions needed to reduce the risks to an acceptable level. In all cases it is essential that the person carrying out the test is competent, having received adequate training and instruction on the risks and the technique to be used to avoid injury. There should always be a sufficiently comprehensive risk assessment carried out, with a clear and unambiguous explanation of the safety precautions. The main techniques that can be employed, most often in combinations, are:

(1)   The use of SELV supplies to the items under test.
(2)   The use of isolating transformers, coupled with the removal of conducting paths to earth in the vicinity of the test area.
(3)   Where isolating transformers cannot be used, shrouding and screening off adjacent live and earthed conductors and providing sensitive earth leakage protection where possible (such as an RCD with a residual operating current of no greater than 30 mA).
(4)   Limiting to a safe level the amount of current that can be supplied into a load representative of the human body. In general this means that the current should be limited to a maximum of 5 mA. As an example, this limitation is commonly applied to the probes of insulation resistance test units which may be energised in excess of 1000 V but which can be

handled without undue risk because the output current is reliably limited to a nominally safe level.

(5)   Carrying out the work in an 'area set aside', delineated by barriers and screens, to prevent both unauthorised access and the tester being disturbed. In some circumstances, such as high voltage test areas, the barriers should be interlocked to prevent access while the unit is energised. People at different work stations on a test bench should be separated by insulating screens.

(6)   The use of warning signs and indicators to highlight the danger.

(7)   The provision of emergency switching devices to ensure the supplies to the test area can be switched off quickly in an emergency.

(8)   Accompaniment of the tester in high risk situations such as high voltage testing of switchgear and other equipment.

(9)   The use of tools and test equipment suitable for the application.

(10)   The wearing of PPE where necessary – such as antiflash clothing where there is a significant risk of high energy flashovers, or insulated gloves to prevent direct contact.

Some examples of these techniques follow.

Earth-free areas

Some types of product, a photocopier perhaps, require adjustments to be made to various components in proximity to exposed live parts as part of the test procedure. If the potential shock voltage exceeds SELV, the danger should be minimised by avoiding the possibility of a shock between a live part and earthed metal or other conducting materials. This is done by carrying out the work in an area that has been made as earth-free as possible and/or by using an isolating transformer to provide an unreferenced earth-free supply. This type of testing should always be preceded by insulation tests to ensure that there is no earth fault which could make the exposed metalwork live. It is also advisable to exclude from the test area conducting materials which might become live from a fortuitous contact with live parts of the product. Thus, wooden or plastic benches are better than metal ones, and plastic rather than metal-cased instruments are preferable. In addition, insulating sheeting such as polythene and neoprene can be used to shroud off conducting paths to earth.

Earthed areas

Some types of electrical apparatus have to be tested and adjusted with their metalwork earthed so an earth has to be introduced into the test area. This would increase the potential shock risk considerably if the normal mains supply, with earthed neutral, was to be employed. One way of avoiding this is to use an isolating transformer and earth each pole of the output winding instead of the neutral through a high resistance in the earthing conductor so that an operator, touching either pole and the earthed metalwork, would experience a potential shock voltage of less than 115 V instead of 230 V and a shock current that is limited by the resistance to a harmless value. The circuit diagram is shown in Fig. 17.13. The relay should be a high impedance sensitive type with an operating current of not more than 1 mA.

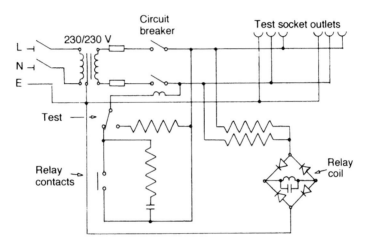

Fig. 17.13 Isolated supply circuit for Class I electronic product testing.

Batch product test

For products that are batch rather than mass produced and supplied in a range of sizes, it is often convenient to establish a test facility between the assembly and packing and despatch departments. Figures 17.14 and 17.15 show an example of a test facility for the transportable refrigeration units used on refrigerated lorries and containers. These units usually have a diesel engine to drive the compressor when in transit by road and a mains voltage electric motor drive for use when parked. The control circuits are usually 12 V or 24 V d.c.

In the factory the units are moved on pallets by forklift trucks. Two sides of the test enclosure comprise push-button controlled barriers which can be raised to permit truck ingress and exit. Limit switches on the barrier pedestals ensure the supply is off when either barrier is raised. There is also an internal safety barrier between the test panel and test piece which has a limit switch on its pedestal to switch off the supply when the tester raises the barrier for

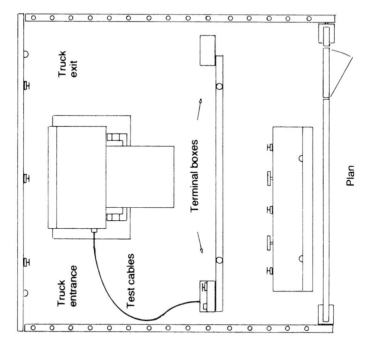

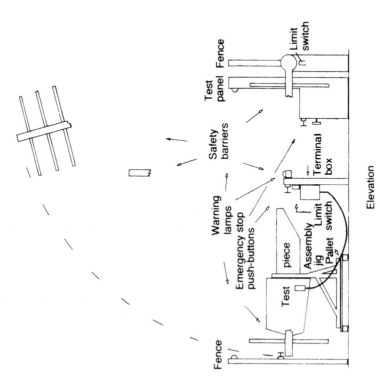

Fig. 17.14 Layout of test facility for batch produced refrigeration units.

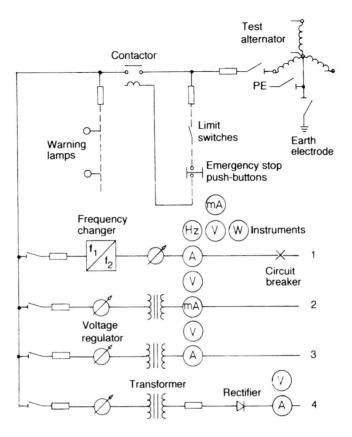

Fig. 17.15 Single line circuit diagram for refrigeration units test.

access to the test piece. Other safety features include connecting the test cables from terminal boxes near the test piece to minimise their length and avoid a tripping hazard at the test panel. The connections between the test panel and the terminal boxes are in floor ducts. The terminal boxes are near the side barriers so that the test cables can be run alongside them and minimise the lengths adjacent to the machine and the tripping hazard. Short lengths of moulded rubber floor mats, with a recess for the cables, could be used over these exposed lengths.

There are a number of manually reset type emergency stop buttons, suitably located, to enable the supply to be cut off, and red warning lamps which are illuminated when the supply is 'on'. To protect the operator against shocks between live parts and earth, the floor of the test area should be non-conducting and the test supply unearthed but with a facility for earthing it if a test specification should require it. When carrying out a.c. tests at voltages exceeding SELV with the supply earthed and the metalwork earthed via the protective conductor, sensitive and rapid operating earth leakage protection may be employed to minimise the shock hazard.

Manually applied test electrodes for HV testing should preferably be used only where the HV current is restricted to a maximum of 5 mA, a potential shock current which is tolerable. However, where it is necessary to burn out faults, for example, and greater currents are needed, it is safer to effect the HV connections to the work piece manually when dead and then control the test at the test panel. As an additional precaution against handling live connections, the operator can be provided with two probes connected together by a short length of flexible conductor and with long insulated handles. By applying these probes to terminals of opposite polarity or between a potentially live terminal and the apparatus metalwork, the operator can verify that the supply is 'off' and the circuit discharged before touching the terminals.

Displayed on the test panel there should be a circuit diagram for the test installation and the test safety rules.

This test layout was devised for a particular range of electromechanical products where bought-in electrical components, which had already been tested by their makers, were incorporated in the product. The testing, therefore, was mainly confined to checking the wiring and the performance of the machine and was different to that required in an electrical manufacturer's factory. For such a manufacturer, of say electric motors and generators, the test layout would have to be suitably modified, but similar safety precautions would be relevant.

High voltage test area

High voltage testing of motors, electrical distribution gear such as circuit breakers and transformers, and generator sets is a specialised activity that carries significant risks. However, the general principles are not much different to those set out for low voltage systems. Figure 17.16 illustrates a high voltage test area of the type commonly found in manufacturers' premises. The fundamental safety principle is that under no circumstances should anybody be inside the test area while the unit under test is energised. It is commonly argued that engineers must be inside the area to read instruments, but this is only permissible if there is no possibility of the person coming into contact with live uninsulated conductors. Moreover, it is well within the state of the art nowadays for instruments to provide remote outputs via data links, without the necessity for people to enter the hazardous area to read them.

There should be a positive indication that the high voltage has been

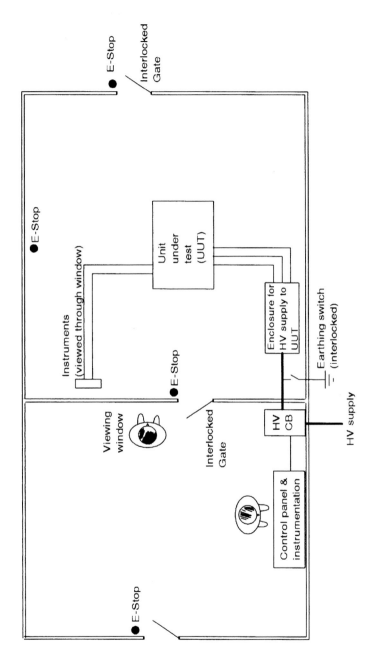

Fig. 17.16 High voltage test area.

removed from the unit under test once the testing is complete and it is advisable to provide automatic earthing facilities on high voltage gear to ensure that all charge is discharged to earth before the people can enter the enclosure.

Chapter 18
Maintenance

INTRODUCTION

An electrical installation which has been properly designed, installed and tested to the requirements of relevant standards, such as BS 7671 in the case of low voltage systems, is likely to be safe and serviceable for some time depending on the nature of its use and misuse. Similarly, electrical equipment that has been designed and built to a relevant standard is likely to be safe at the time of its supply. The result is that some managements, particularly of smaller enterprises, who are preoccupied with running their business tend to ignore the need for maintenance and may well do nothing about it until the consequence of this neglect appears in the form of a breakdown, accident or fire. Those who have thought about it often dismiss maintenance as an unproductive cost with no return on the investment so spend as little as possible, doing nothing until there is an incident. In taking this all-too-common approach, they lose sight both of their legal obligation to maintain systems so as to prevent danger and the economic advantages of avoiding incidents; some careful consideration of these two aspects may induce them to inaugurate a planned maintenance system.

STATUTORY REQUIREMENTS

In all locations where people are at work, except servants working in domestic premises, maintenance of the electrical installations and systems is a statutory requirement covered by Regulation 4(2) of the Electricity at Work Regulations 1989. In addition to this, Regulation 5 of the Provision and Use of Work Equipment Regulations 1998 requires work equipment to be maintained in an efficient state, in efficient working order and in good repair; this requirement for maintenance embraces electrical equipment and apparatus and electrotechnical control systems on machinery and plant. Moreover, irrespective of whether or not work is being done, the Electricity

Supply Regulations, Regulation 17, require an electricity supplier's works to be 'constructed, installed, protected (both electrically and mechanically), used and maintained as to prevent danger ... so far as is reasonably practicable'.

These Regulations require that maintenance is carried out to prevent danger. This means that some form of inspection, test and repair has to be done in advance so as to anticipate dangerous faults and to ensure that they do not occur, so far as is reasonably practicable. It is evident that neglecting an installation or equipment until an incident occurs and then taking remedial action does not constitute preventive maintenance and does not comply with the law.

DEFINITION AND FORMS OF MAINTENANCE

A planned maintenance system comprises a procedure for the inspection, test and repair of the electrical installation and connected apparatus to avoid breakdowns and dangerous faults. It is most commonly regarded as a preventive surveillance system whereby apparatus is inspected and if necessary tested and repaired before it is likely to break down or develop a dangerous fault. The relative contributions of visual examinations and testing have already been considered elsewhere in this book.

In reality, there are a number of different forms that maintenance can take. Perhaps the most common is routine preventive maintenance of the type prescribed by BS 7671, where periodic inspections and tests are carried out to try to detect any failures or degradation that may lead to danger before that danger is realised. Whereas this is an effective technique it does have the disadvantage that it requires intervention into the equipment on a periodic basis regardless of the actual condition of the equipment or the likelihood that a fault has occurred or the likelihood that the characteristics of the equipment may have degraded. This means that equipment that may be in perfectly good condition has to be dismantled, with the possibility that the intervention itself will cause failures or degradation. In the light of this, there have been attempts over the years to devise other maintenance strategies that take into account the actual condition of the equipment and/or its reliability.

One important development in this area has been that of 'condition-based maintenance' (CBM). The objective of CBM is to use non-invasive techniques to detect the current state of systems both to decide whether or not intervention is needed and accurately to predict their remaining useful lives. This enables engineers to perform maintenance only when needed to prevent operational deficiencies or failures, essentially eliminating costly periodic preventive maintenance and reducing the likelihood of plant failures. CBM uses a variety of sensor systems to detect and diagnose emerging equipment problems and to predict how long the equipment can effectively serve its operational purpose. The output of the procedures is maintenance personnel being alerted to developing problems, enabling maintenance activities to be scheduled and performed, as needed, before operational effectiveness or safety are compromised. The technique is applicable across a broad range of technologies but, in the electrical field, techniques such as partial discharge testing and thermographic imaging can be used:

- Partial discharge testing makes use of the fact that in high voltage systems any voids in insulation will usually fill with gas. The gas will ionise, and discharge activity caused by the high electric field strengths will take place. This discharge activity can be detected by radio frequency monitors, with the extent of the detected activity giving an indication both of the presence of voids and the likelihood of an insulation failure. This means that corrective action can be taken before a catastrophic insulation failure occurs. The technique is used on high voltage distribution boards to monitor the likes of compound-filled busbar chambers and cable boxes and cable joints.
- Thermographic imaging uses the fact that high resistance points in conductors, such as may occur at weak and loose connections, or locations where insulation has been damaged and there is leakage current between conductors at different voltages, will generate heat. This heat can be detected using cameras that produce images in the infrared band, allowing 'hot-spots' to be detected and corrective action to be taken before a catastrophic failure occurs. The technique is increasingly being used to monitor both high and low voltage distribution boards and other electrical apparatus such as motors, especially in circumstances where a need for continuous operation means that the equipment cannot be switched off for invasive testing. Typical examples of this are refrigeration and freezer units in 24-hour supermarkets, and IT and power distribution systems supporting 24-hour banking operations. Care has to be taken in its implementation because covers frequently have to be removed to expose the equipment inside enclosures and cabinets – this may lead to the imaging work having to be treated as 'live work' in the context of Regulation 14 of the Electricity at Work Regulations, in which case precautions will need to be taken against the possibility of the person using the camera making direct contact with exposed live conductors.

It is worth observing that these techniques, on their own, do not provide a full 'health check' of an electrical system. Thermographic imaging, for example, does not give an indication of the continuity of earth conductors or the value of earth loop impedance on protective systems – it cannot, there-fore, give an indication of the condition of the systems installed for earthed equipotential bonding and automatic disconnection protection against indirect contact injuries. They should therefore be treated as being just one element of the overall armoury of maintenance techniques available.

Another important concept in preventive maintenance is that of reliability centred maintenance (RCM). RCM is a systematic process of preserving a system's function by selecting and applying a range of preventive main-tenance techniques. It differs from most approaches to preventive main-tenance by focusing on function rather than equipment. In general, the concept of RCM is applicable in large and complex systems such as chemical plants, oil refineries and power stations.

The RCM approach arose in the late 1960s and early 1970s when the increasing complexity of systems (and consequent increasing size and cost of the preventive maintenance task) forced a rethink of maintenance policies among manufacturers and operators of large passenger aircraft. Pioneering work on the subject was done by United Airlines in the USA in the 1970s to support the development and licensing of the Boeing 747. The principles which define and characterise RCM are:

- a focus on the preservation of system function;
- the identification of specific failure modes to define loss of function or functional failure;
- the prioritisation of the importance of the failure modes, because not all functions or functional failures have equal consequence; and
- the identification of effective and applicable preventive maintenance tasks for the appropriate failure modes.

It will be appreciated from this that there is a diverse range of techniques and methods that can be applied to the maintenance of electrical systems. In simple and non-complex systems, the normal approach will be to adopt the simple planned preventive schemes advocated by BS 7671 and similar stan-dards. In more complex systems, the innovative techniques associated with schemes such as CBM and RCM should be considered, although specialist advice will frequently be needed if informed judgements are to be made.

The following sections largely assume that a simple planned preventive scheme is appropriate.

THE PLANNER

A competent person familiar with the installation and the connected equipment, and their usage, is best placed to devise the planned maintenance system. For larger establishments with a maintenance department, the senior electrical person should do it. For smaller premises with no in-house maintenance staff or person with the appropriate competencies, the services of a consultant or similar entity should be sought. As an alternative, the electrical contractor appointed to do the maintenance work may be able and willing to do the planning as a separate item of contracted work.

THE MAINTENANCE SCHEME

Circuit and apparatus identification

If not already available, a circuit diagram or schematic of the installation should be prepared, and the circuits and connected apparatus on it should be identified. The numbers and other identifying labels marked on circuits, switchgear, control gear and apparatus should match those in the drawings and schematics. Portable and transportable apparatus, which is not usually shown on a circuit diagram, should also be marked and recorded. Existing circuit diagrams should be checked against the installation and amended if necessary to reflect any alterations.

Register

Next, a register of the circuits and apparatus should be prepared. It will probably be most convenient if this takes the form of a loose-leaf ledger or computer database or spreadsheet. In larger premises it will be worth considering investing in one of the many specialist software packages available to support preventive maintenance schemes. This will facilitate adding new items, removing those that are redundant and adding additional records for any items as required. For each item there should be a description which identifies it and its location and gives essential information about it, such as the name and address of the maker; cost; location of the manufacturer's installation, use and maintenance instructions; and a list and location of spare parts. There should also be a maintenance schedule showing the frequency of inspections and what inspection, testing, adjusting and renewal of parts should be done on each occasion.

Additional pages should bear a checklist for the maintainer to tick the items done and to provide space for remarks. Other pages should be provided to record defects and breakdowns and the rectifying action that has been taken.

Small installations

The scheme can be simplified for small installations, such as most domestic accommodation, small shops and offices with only a few circuits for a cooker, water heater, lighting and socket outlets fed from a consumer unit. For instance, if a circuit diagram is not available, a circuit schedule may be used and the inspections and tests carried out as recommended by BS 7671 at intervals of five years and ten years for domestic installations. Portable and transportable equipment should be routinely inspected, with hand-held portable items being tested at a suitable frequency.

The 'events diary'

The inspection dates should be entered in a diary or call-forward system or planned preventive maintenance software package which is consulted by the supervisor when planning the work schedule.

Frequency of inspections

The average installation

Except for those installations where there is a mandatory requirement, such as petrol filling stations, the frequency of inspection is a matter of judgement by the planner based on experience and published guidance. For the fixed installation, i.e. excluding the connected equipment, the IEE gives a list of recommended inspection intervals in Table 4A of Guidance Note No. 3 – Inspection and testing. The aim should be to choose the interval between inspections of an item so that it is checked just before it is likely to become defective. There is no virtue in increasing the frequency of inspections; unnecessary dismantling may itself cause trouble by damaging parts such as gaskets. The interval lengths should be reviewed periodically and adjusted in the light of experience, based on changes in use and the breakdown and defect reports. The makers' recommendations are likely to be conservative. The intervals given by them may sometimes be safely extended, but only after the guarantee period has expired to avoid invalidating it.

The failure rate/time curve for most items is shaped like a bath, the failure rate being plotted vertically and the time horizontally. The higher failure rate soon after commissioning is due to assembly errors and premature failure of defective components. For the more important items, at least, it is prudent to increase the number of inspections during this time. After this teething period, the failure rate should fall to a minimum until the apparatus nears the end of its economically useful life when general wear and fatigue stresses cause the failure rate to increase again.

Equipment

Electrical equipment needs to be subjected to inspection and testing just as much as electrical installations. In recent years there has been a tendency for employers to overtest their portable appliances but ignore the need to inspect and test fixed equipment and the installation. A source of authoritative advice on the form that inspection and testing of equipment should take, and the appropriate inspection and test intervals, is the IEE's Code of practice for in-service inspection and testing of electrical equipment. The second edition of this was published in 2001. This publication quite correctly stresses the need for before use and regular visual examinations of equipment because most faults, such as scuffed/abraded insulation and cracked plugs, can be found simply through inspecting the equipment. It also notes that in some benign environments, such as offices and hotels, there may well be no requirement to test items such as Class II equipment at all. This would not apply to equipment used in harsher environments such as construction sites, where regular combined inspections and tests are recommended (see Chapter 11).

Note that some electrical equipment associated with machinery such as lifting equipment (lifts, cranes and mobile elevating work platforms) and power presses, is subject to statutory inspections and thorough examinations under legislation such as the Lifting Operations and Lifting Equipment Regulations (LOLER) 1998 and the Provision and Use of Work Equipment Regulations 1998.

Special installations

In some cases, the maintenance has to be done at particular times. Continuous process plants are an example. They are run for prolonged periods and then shut down only at predetermined intervals for maintenance and modification/upgrade purposes. Electrical components, vital to the operation of the plant, usually merit inspection on each such occasions and any parts, subject to wear deterioration, should be replaced prematurely to insure against failure operation. Brushes, contactor contacts and the oil of oil-immersed switchgear and control gear, are examples.

Construction sites

Construction sites are a special case. The electrical installation is temporary and is altered from time to time to meet the changing needs of the construction programme. The installation is subject to adverse environmental conditions and rough handling. More frequent inspections are needed therefore, to ensure safety. IEE Guidance Note No. 3 and HSE's Guidance Note HS(G)141 both recommend intervals not exceeding three months for the electrical installation. HSG141 also contains recommendations for the intervals at which equipment such as handtools, lifts and extension leads should be inspected and tested.

Maintenance instructions

For most items, the maintenance engineer should have little difficulty listing the work to be done when the item is inspected. Among the literature available, BS 6423 and BS 6626 give excellent advice on LV and switchgear and control gear maintenance. For the more complex apparatus the makers' instructions should be obtained and followed. The aforementioned HSE and IEE guidance is also helpful.

Tools, special equipment and instruments

When compiling the register, the planner should have regard to the equipment needed to service particular items of apparatus, should ensure that it is available, and note it on the service schedule so that the maintainer will take it when going to inspect the relevant apparatus. For live working, insulated tools should be provided together with insulating screens, gloves and mats.

There must also be a range of instruments including insulation testers, multimeters, potential indicators, loop impedance testers, RCD testers, current transformers, wattmeters and perhaps relay testing equipment. Expensive instruments, such as oscilloscopes, which may be only occasionally required and which may require expert handling are probably not worth buying, but can be hired when needed or can be provided by a specialist called in to deal with a problem that cannot be tackled in-house.

OPERATION OF THE SCHEME

Introducing the inspection programme

In the case of a new installation, it should not be too difficult to ensure that inspections of each item are carried out on, or about, the due date because teething-trouble breakdowns and minor alterations to the installation should not be so time consuming as to disrupt the inspection programme seriously.

It is usually much more difficult to introduce an inspection programme in premises where it has not previously existed and where the maintenance staff are fully stretched coping with successive breakdowns which occur because of this neglect. One way out of this dilemma is to augment the staff temporarily so that some of them can concentrate on the inspections. As more and more of the equipment is being properly serviced, the number of breakdowns should diminish until such time as the scheme is fully operational and the extra staff are no longer needed.

It may not be easy to find temporary staff who are conscientious, qualified and suitably experienced. It may, therefore, be better to borrow them from a local electrical contractor, preferably one who put in the original installation or who has been employed for alterations or extensions to the installation and has some knowledge of it. Alternatively, an employment agency which specialises in the provision of temporary tradesmen may be able to help.

Disruptions

Extensions and alterations to the installation, breakdowns and staff shortages all tend to disrupt the inspection programme. If the inspections are late, the breakdown rate is likely to increase which further disrupts the programme and may induce a snowball effect, culminating eventually in the staff's time being solely taken up with breakdown repairs. It is essential to avoid such an eventuality, so when new work or alterations to the installation are required, they should only be carried out by the maintenance staff if it can be done without serious interference with the maintenance programme. If there is any doubt, the work should be placed with an electrical contractor.

There should be a prioritisation system to determine which inspections must done on time and which may be deferred. There should, however, be a safeguard to ensure the deferments are of limited duration to avoid an item's inspection being so late that it becomes dangerously unserviceable or breaks down.

Reporting defects

Between inspections, some equipment will become defective and may become dangerous so the users should be encouraged to look for and report such defects as soon as they are aware of them, preferably as part of a pre-use inspection. They should cease to use the equipment until the defect has been repaired. Portable apparatus with flexible cables and plugs is more likely to be damaged than fixed equipment. It is also more likely to be hazardous to the user, who should be reminded periodically of the danger and the need to report incipient defects before they become dangerous.

Training

The success of any planned maintenance scheme is dependent on the skill of the staff and its conscientious application. Apart from the selection of competent people to carry out the work, in-house training is required to familiarise them with the installation and the planned maintenance scheme, and hopefully to inspire them with some enthusiasm for the task. This might be done by explaining the importance of preventive maintenance, their role in it and its relevance to the safety of the staff and economic success of the business. A training course might include:

■ The safety rules and procedures in force in the premises, including safe isolation practices (including the use of permits to work and other safety documents where appropriate) and safe systems of work for live working.
■ A brief history of the enterprise from its formation to the present time, and its current objectives.
■ The circuit diagram(s) for the installation from the power source(s) to the connected equipment.
■ The reasons for a planned maintenance scheme, including its economic justification.
■ A detailed explanation of the scheme in force.
■ The task of the maintenance electrician including:
  □ liaison with the production staff to arrange for the shutdown of the equipment for inspection and to obtain details of any defects they have observed;
  □ the use of the checklist;
  □ the use of the makers' instructions;
  □ what to enter under 'Remarks' on the report sheet;
  □ investigating the cause of failures;
  □ dealing with insoluble problems by seeking help from the supervisor rather than guessing the answer;
  □ looking for trouble, i.e. using one's senses of sight, smell and hearing, when moving around in the premises, to detect possible defects. For example, be on the lookout for damage to flexible cords and cables, the odour of burnt insulation and the sound of a noisy motor bearing.

From time to time, the initial training should be supplemented to keep the staff informed about any problems that arise, new apparatus, changes in practice, etc. Such meetings should be of the discussion type to encourage full participation by all staff members.

Good housekeeping

Apparatus should be kept clean and the legal requirement for it to be accessible must be observed. Dirt and obstructions can hide defects and hamper the task of the maintenance staff who should not have to move obstructions to obtain access to defective apparatus or spend time cleaning it – time which might be more usefully employed in repairing the defect.

Selection of equipment

The engineer responsible for the maintenance should also be consulted before new equipment is purchased so that advice can be obtained on its maintainability. Good quality apparatus, designed for easy maintenance, may prove to be more economical than cheaper apparatus of lesser quality which is more difficult to maintain. The engineer should also be familiar with the environmental conditions and can advise on any protective features required, e.g. whether the apparatus needs to be dust-tight, weatherproof, or suitable for use in a hot location.

Siting of new equipment

Again, the maintenance engineer should be consulted to ensure that the equipment is accessible and there is adequate space and, where necessary, lifting provisions to facilitate dismantling for maintenance. The siting should have regard to the proximity of other apparatus, the operation of which too close may endanger the safety of the maintenance staff.

Specialist help

Some modern equipment is complex and it may well prove to be too difficult for the ordinary electrician to find a fault and repair it. For example, if the photo-electric guard on a power press fails, the electrician can check the relevant limit switches and rectify a fault in them and can replace burnt-out lamps, but if the fault is in the electronic circuits, it will be necessary to call an electronics engineer, preferably from the maker. Unskilled tinkering with the electronic circuits or bypassing them to get the press back into production could have serious consequences for the press operator whose safety is dependent on the correct functioning of the guard. Where there is sufficient electronic apparatus to justify it, a specialist technician could be employed but otherwise specialist help should be sought.

Not all electricians are capable of servicing HV equipment. Where it is confined to substation switchgear, for example, and perhaps a few large electric drives, it may be convenient to arrange for its maintenance by a supply company rather than entrust it to less experienced in-house staff. It is crucial that HV work should only be carried out by competent personnel who have the appropriate skills, knowledge and experience, and who have been authorised to carry out the work.

Servicing and adjusting the relays of protective gear is another item that could be contracted out to the supply company or other specialist contractor with the appropriate equipment and expertise. Major repairs, such as rewinding motors, are best carried out by electrical repair specialists. The periodic calibration of instruments is a matter for the makers or a metrology laboratory.

Chapter 19
Checking Small Domestic Electrical Installations

INTRODUCTION

There is often a demand for reports on the condition of tenants' electrical installations in rented accommodation. This arises because some landlords fail to maintain the accommodation in a safe and serviceable condition and ignore their tenants' complaints. In many circumstances such as this, a tenant will approach a local authority or solicitor for assistance in resolving the matter, and this will frequently result in a building surveyor or an environmental health consultant being asked to inspect the accommodation and to produce a report. If the tenant indicates that he or she has had electrical problems or suspects that the electrical installation is faulty, an electrical engineer will often be asked to inspect and test the installation and submit a report.

This chapter suggests a procedure that can be adopted by electrical engineers who are asked to carry out this type of work.

LEGISLATION

Legislation dealing with landlord and tenant relationships does not have any specific electrical safety requirements; however, if landlords provide an electrical installation as part of the rented accommodation (which will invariably be the case), they have a responsibility to maintain it in a safe and serviceable state. Failure to do so constitutes negligence, which is actionable.

In the case of houses in multiple occupancy, the Housing (Management of Houses in Multiple Occupation) Regulations 1990 stipulate that a landlord has a duty to ensure that the electrical installation in the common and accommodation areas is in and is maintained in a state of repair and proper working order. It is a moot point whether this can be interpreted as requiring the installation to be safe. Local authorities who license such houses will usually require evidence to be produced to show that this duty is being satisfied before they will issue licences.

337

As there is no electrical legislation for domestic premises, the safety criteria have to be the requirements of the non-statutory BS 7671 : 2001 (see Chapter 10). This is the recognised safety standard observed by all reputable electrical contractors and as such is generally accepted by the courts as the benchmark standard. Compliance with the standard is invariably a requirement in specifications for electrical installation projects. Non-compliance usually means that the installation is likely to be unsafe, with the attendant risks of electric shock and/or burn or fire. However, it should be borne in mind that amendments to standards such as BS 7671 are not meant to be applied retrospectively. So the fact that an installation completed, say, 20 years ago does not comply with the latest edition of the standard should not be taken as an indication that the installation is unsafe if it complies with the edition of the standard extant at the time the installation was completed. The acid test is not so much whether the installation complies with a standard but whether or not it is safe, and professional judgement will frequently be needed to determine this. Non-compliances with the current edition of the standard should be recorded, but the safety consequences of the non-compliances should also be considered and noted.

PROCEDURE

Papers

The engineer charged with investigating the complaint should obtain as much detail of it as possible. A copy of any statements made by the tenant and other potential witnesses and a copy of the surveyor's report, if one exists, are good starting points. The latter can be useful as, apart from any electrical comments, it will probably reveal wetness problems from leaking roofs or pipes, condensation, faulty damp courses, flooding, etc. which may have an adverse effect on the electrical installation.

Visit

A visit should be made during daylight hours as the insulation and continuity testing involves switching off at the consumer unit, so there will be no lighting. It is advisable for the tenant to be there to provide first-hand information on the length of tenure and the electrical complaints.

If the complaint leads to litigation, it is improbable that the judge and counsel will visit the premises, but they are most likely to rely on the report when arguing the case. Therefore, an outline plan showing the layout and location of the electrical points, supported by photographs of any defects, is

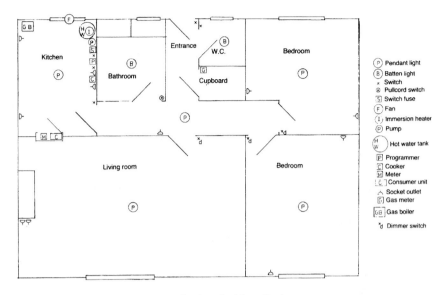

Fig. 19.1 A typical small domestic electrical installation.

helpful and may enable them to understand it better, particularly as it is probable that they will have little electrical knowledge. Figure 19.1 is an example of a typical installation layout sketch.

When going round the premises with the occupier to prepare the layout sketch, defective items and infringements of the standards should be noted. Socket outlets' polarity should be checked with a neon indicator which will also show whether the socket outlet has an earth connection. Also check the main equipotential bonding of the water service, gas installation and other service metal pipes and also of the central heating, air conditioning systems, exposed metallic structural parts of the building and the lightning protection installation, if any. In the bathroom, check that the supplementary equipotential bonding has been properly done and that the bathroom installation complies with section 601 of BS 7671.

The intake

A normal modern intake into a house comprises a CNE concentric service cable terminating in a plastic sealing box. The supply company provides an earthing terminal connected to the cable sheath, which is the CNE conductor, and a cut-out consisting usually of a 60 A or 100 A BS 1361 fuse. The meter tails are generally PVC insulated and sheathed conductors. The Electricity Supply Regulations require them to be marked for polarity. Quite often the meter tail wiring is untidy and not secured so as to relieve its terminations of strain.

Note that the Electricity Supply Regulations require the earthing conductor to comply with the table in Regulation 7, which means a minimum cross sectional area of $10\,mm^2$ if the service cable does not exceed $35\,mm^2$. This size also applies to the main equipotential bonding conductors.

Older installations usually have an armoured paper insulated and lead sheathed service cable providing an SNE supply. The consumer's earthing terminal should be connected to the lead sheath of this service cable. Look out for earthing to the incoming water service pipe, which is no longer permitted. Check that the size of the earthing conductor complies with BS 7671, section 543, or, if the consumer has his or her own buried earth electrode, is in accordance with both sections 542 and 543.

In rural areas there are still some TT supply systems where consumers have to supply their own earth, in which case it is usually necessary to protect the installation by means of a current-operated earth leakage circuit breaker because of the difficulty of providing an earth of sufficiently low resistance to ensure that the protection operates within the prescribed time in the event of an earth fault. Note that voltage-operated earth leakage circuit breakers are no longer acceptable.

Check the position of the meter. The Electricity Supply Regulations require it to be installed in the consumer's premises unless it is more reasonable for it to be located elsewhere, in which case it should be accessible to the consumer. This is more likely to occur in blocks of flats and housing in multiple occupation. In such premises the supply authority may bring in only one service cable on to a busbar and then feed each tenant via cables in conduit or sometimes in MIMS cable. The latter is not a good choice because the insulation is hygroscopic and if the end seals fail, the cable can break down. It may also fail if subject to HV transients, so such cables should be fitted with suppressors. If the conduit or sheath of the MIMS cable is used as the earthing conductor, check that the connections are satisfactory and of low resistance and any joints in the conduit provide adequate continuity.

As service cables have no overload or short circuit protection, a fault will cause arcing until it burns itself out, so there is a potential fire risk although such breakdowns are uncommon. Nevertheless, where such cables are inside the premises, it is worthwhile checking that they are so located or shielded that the arcing would not ignite anything else and start a fire; the Electricity Supply Regulations, Regulation 17, refers. Where the supply is from an overhead line, there is a small potential fire risk from an HV surge caused by a lightning strike or near strike on the line conductors. Again, supply companies do not protect their service cables against lightning, so suppressors should be fitted where connected equipment is not protected against such transients and/or where there is a history of high thunderstorm activity.

The consumer's installation

Distribution board

Post-World War II installations will usually have a consumer unit, which is a small distribution board with a double pole isolating switch and a phase busbar to feed the fuses or MCBs controlling the final circuits. There will be terminal blocks for the neutrals and protective conductors. The earth terminal block may be utilised as the consumer's earthing terminal. Some of the older models may have wood frames and/or may be backless. As BS 7671 requires connections to be made in non-flammable enclosures, the wooden framed type do not comply and the backless ones are acceptable only if mounted on non-flammable material. If rewirable fuses to BS 3036 are used, check that the correct size of fuse wire has been used. For cartridge fuses, check that blown fuses have not been 'repaired' with a bit of fuse wire spanning the contacts or the cartridge replaced by a nail, hairpin or the like. There should be only one final circuit connected to each fuse carrier or MCB.

The circuits should be identified on a diagram or schedule and there should be separate circuits for lighting, socket outlets, cooker, immersion heater and central heating. Although not mandatory, it will usually be found desirable and convenient to have a separate socket outlet ring main for the kitchen to cater for the substantial load of the appliances therein, a number of which may operate at the same time, and it is also convenient to have a separate socket outlet ring main for the socket outlets on each floor. Again, it is desirable to have several separate lighting circuits to avoid a fault on one plunging the whole house into darkness. For protection against electric shock, consumer units are often of the split variety with a residual current circuit breaker (RCCB) providing earth leakage protection on the socket outlet circuits.

Some pre-World War II installations may be found with an ironclad splitter switchfuse usually with four or six rewirable fuses in porcelain carriers, both phase and neutral being fused. This is, of course, no longer permitted. It stems from the time when most small dwellings had few electrical appliances and electricity was mainly used for fixed lighting. Sometimes there were a few BS 546 2 A and 5 A socket outlets to supply table and standard lamps and perhaps a vacuum cleaner. Subsequently, more appliances were introduced and the BS 546 outlets were replaced by BS 1363 13 A socket outlets by the occupier or a 'cowboy' electrician without altering the distribution system. So the socket outlet wiring will probably be found to be of inadequate size for the potential load and oversized fuse wire may be found in the fuse carriers. Alternatively, there may be a double pole switch or switchfuse controlling a small distribution board which may be wooden-cased and with a glazed front. It is unacceptable because the wood is flammable.

Wiring

The ends of the outgoing circuits at the consumer unit will indicate the type of wiring employed. If these ends are exposed outside the fuseboard they must be secured to relieve their terminations of strain, and cable sheaths should terminate inside the enclosure. Entry holes should be bushed. Look out for VRI lead-covered cables. There was a vogue for these between World War 1 and II. The lead sheath was usually used as the protective conductor and continuity was a problem because the clips used on the sheath for the purpose loosened, owing to the cold flow properties of lead, and corroded. Impurities in the lead and sulphur from the vulcanised rubber could cause perforations in the lead sheaths and, as it aged, the rubber perished. So any such wiring found will be unserviceable and should be discarded. VRI conductors were also used with tough rubber sheaths. With age the rubber hardens and cracks and if disturbed will part company with the conductor, thereby no longer being serviceable. VRI cables were also made with a tape binding over the rubber and then a compounded textile braid was applied over the tape. These cables were installed in metal conduit or ducting and sometimes in wood capping and casing. This type of enclosure is no longer permissible because it is flammable.

VRI and TRS cables were superseded by PVC insulated and PVC insulated and sheathed cables in the decade after World War II. Any TRS or VRI cables now found will almost certainly have perished insulation, will no longer be serviceable and will need replacing. Screwed steel conduit may be utilised as the protective conductor, but light gauge conduit, too thin to be screwed or with close joints connected by pin grips or slip jointed, cannot be relied on for continuity and is no longer allowed by BS 7671. There were also clamped joint versions which, although not as good as screwed steel conduit, may provide adequate earth continuity, which is ascertainable by testing. Since the introduction of PVC insulation, most domestic wiring has been in two core and earth PVC insulated and sheathed cables concealed in the structure and often without further protection. Check for the following infringements of BS 7671:

■ That the bare protective conductor is not sleeved at terminations where it emerges from the sheath.

- Under floorboards where the cable traverses joists, it has to be at least 50 mm from the floorboards to avoid damage from fixing nails or screws.
- Mechanically unprotected cables concealed in walls or partitions at a depth of less than 50 mm have to be run within 150 mm of the top of the wall or partition or within 150 mm of the angle formed by two adjoining walls or partitions. Where the cable is connected to a point outside these zones, it has to be run in a line either vertically or horizontally. This can be checked with a metal detector.
- There should be no loose wiring. Cables should be secured except where normally inaccessible, such as in roof spaces.
- Any polystyrene thermal insulation found should not be in contact with the cables as PVC is adversely affected by it.
- Cable sizes should be correct.
- Where cables are run in thermally insulated spaces or in hot locations, ensure that the conductor size is adequate when derated.
- Ensure that the protective conductor is available at every point.
- Where cables pass through fire barriers, check that the opening is sealed.
- Ensure that segregation of the telecommunication, fire alarm, emergency lighting and extra-low voltage circuits from the rest of the installation circuits has been effected.
- Cables are not subjected to condensation dripping from metal sinks, cold water pipes and tanks or made wet from leaks.
- In damp locations and where condensation occurs, check that Class II accessories such as switches and socket outlets are watertight or otherwise designed to prevent a film of moisture providing a current path from a 'live' part inside to the exterior where it may be touched.
- Ensure that there is adequate separation between cables and non-electrical services such as gas installation pipes and equipment so that any work on one service does not adversely affect the other.

Main equipotential bonding

The main equipotential bonding cable sizes are given in BS 7671. For most dwellings having a service cable neutral conductor not exceeding $35\,\text{mm}^2$ and where the supply is PME, the minimum size is $10\,\text{mm}^2$. Otherwise it should be not less than half the size of the earthing conductor, with a minimum of $6\,\text{mm}^2$. In practice, $6\,\text{mm}^2$ is generally adequate for non-PME supplies. If separate conductors are not used for each point, then a conductor common to several points must be continuous so that it is not broken if detached at one point. Check that all connections are accessible for periodic testing, have the safety label and are not corroded. Connections under sinks are most likely to be affected by leaks or dripping condensation.

Rooms containing a bath or shower basin

For rooms containing a bath or shower basin, check for compliance with section 601 of BS 7671. Note that there must be no surface wiring in a metallic enclosure, and heaters with touchable elements must be out of reach of anyone using a bath or shower. The only socket outlets allowable are for shavers fed from BS 3535 transformers. The only permissible switches and electrical controls touchable by a person using a bath or shower are those for water heaters complying with BS 3676. Exposed lampholders within 2.5 m of a bath or shower cubicle have to have a BS 5042 safety shield.

It is important to check that the supplementary equipotential bonding has been properly done. This entails connecting together all simultaneously accessible exposed conductive and extraneous conductive parts by means of an insulated conductor. In effect this means, for example, connecting a metal bath to its hot and cold water and waste pipes. The pipework itself is not regarded as an equipotential conductor for this purpose. Figure 10.2 shows the connections required. As it is unlikely that mechanical protection will be provided for these bonding conductors, the minimum size will be $4 \, mm^2$ in most cases. Although most bathrooms will have wooden floors, bathrooms in flats, basements or on ground floors may have solid floors which are conductive and earthed. The standards do not require such floors to have equipotentially bonded grids, except where electric heating is embedded in the floor. For safety, it is suggested that such floors should have an insulating cover when the bath is in use.

Socket outlets

Sockets should not be mounted on skirting boards where they are vulnerable to damage from floor cleaners and the like, but should be at a sufficient height above the floor or worktop to enable the plug to be inserted without unduly bending the flexible cord. The connections should be made in a non-inflammable enclosure which usually means a sunken steel box for flush socket outlets or a plastic or steel one for the surface type. The sheaths of sheathed cables should terminate in the box and, for surface socket outlets where the wiring is on the surface, the cables should be secured to relieve the terminations of strain. Box entry holes should be bushed and bare protective conductors sleeved. If conduit is used as the protective conductor, the earthing terminal of the socket outlet has to be connected to the box earthing terminal by means of a separate conductor. Check whether there are socket outlets outside the equipotential zone, such as in the garden, or whether there are sockets which will foreseeably be used to supply equipment outside the equipotential zone. These sockets have to be protected additionally by means of a sensitive RCD with a rated residual operating current not exceeding 30 mA. If they are exposed, they should be weatherproof.

Switches

Although most switches will be Class II, the protective conductor should be available at the point. Again, the connections should be made within non-flammable enclosures, the sheaths of sheathed cables terminated inside the enclosures, bare protective conductors sleeved and for surface wiring the cables secured to relieve their terminations of strain. Check whether the incoming phase terminal has been used for looping instead of at the ceiling rose. This practice is not acceptable for steel conduit systems as the currents will not be in balance and the magnetic effect will cause a circulating current in the conduit.

Flexible cables

Flexible cables are often used to supply extractor fans, immersion heaters, pumps and controls for the central heating and domestic hot water systems. They are subject to the same requirements as for fixed wiring and where accessible should be secured so as to relieve their terminations of strain. Check the condition of lighting pendants which may have deteriorated from age or been damaged by a hot lampholder.

Downlighters

Ceiling recessed downlighters have caused fires. Samples should be removed to check whether there is adequate ventilation space in the ceiling void to dissipate the lamp heat, particularly where there is thermal insulation. Check also that the cables are of adequate size where the luminaire does not have an integral transformer and the extra-low voltage lamps are fed from a separate transformer.

Testing

It is not necessary to carry out all the tests prescribed by BS 7671, which should have been done when the installation was new. Sufficient tests are required, however, to prove the safety or otherwise of the installation now. The method of testing is described in detail in Chapter 17. The recommended tests are:

(1) *Continuity*. The continuity of the protective conductors in sample circuits should be checked with an ohmmeter. It is not necessary to test

the phase and neutral conductors as any break in them will be self-evident, but all the conductors in the socket outlet ring circuits should be checked as any break could lead to overloading of the live conductors and a failure of the protective conductor under fault conditions. After testing the bonding conductors, it is worth applying the test again to the protected item after reconnecting the conductor. A higher reading indicates a defective joint.

(2)   *Insulation.* The tests described in Chapter 17 can be curtailed and reduced to:

    (a)   Unplug portable appliances and isolate fixed equipment as it is the wiring installation only that is being tested. Open the main switch. Connect the phase busbar to the neutral terminal block. Disconnect or short out any suppressors, electronic apparatus such as a programmer, or lighting dimmer switches, but close other lighting switches. Then apply a 500 V d.c. insulation test between the phase and neutral conductors and the earth terminal block. If the reading is low, remove the temporary connection between the phase busbar and neutral terminal block, open all lighting switches and test between the phase busbar and earth terminal block to determine whether the fault is on the phase or neutral conductors. If on the phase, test each circuit by removing the fuses or opening the MCBs in turn. If on the neutral conductors, each circuit will have to be tested by disconnecting them in turn from the neutral terminal block.

    (b)   It is not usually necessary to test between the phase and neutral, but if it is done, electronic equipment will need to be disconnected or shorted out and one pole of the bell transformer, indicator lamps and capacitors disconnected and lighting switches opened. A modern installation usually has an insulation value of over 20 megohms. A reading of only 0.5 megohms is required, but low readings of the order of 2 megohms should be investigated.

(3)   *Earthing.* The connection to earth has to be of sufficiently low resistance for the protective devices to operate within the prescribed time in the event of an earth fault. To test this, an earth loop impedance test should be done. In most cases, a test at the socket outlet furthest from the consumer unit is sufficient. On TN systems where the supply company provide an earth, readings of 0.3 to 0.5 ohms are usual and sufficiently low for the purpose. On TT systems, where consumers provide their own earth, higher readings are likely and overcurrent protection for earth faults may not be adequate, in which case earth leakage protection is required by means of an RCCB. The device can be tested for operation by pressing the test button, but for operating time, test with

the appropriate instrument. An RCCB is often fitted on consumer units for protection against electric shock. It is usually to control the socket outlet circuits. To do the earth loop impedance test it will have to be bypassed.

Tenant's complaints

Apart from the obvious defects, such as damaged or unserviceable points, sometimes there are intermittent faults which are more of a problem, such as the occasional rupturing of a particular fuse or the tripping of one of the MCBs. Questioning the occupier about the circumstances of such outages will often provide answers which point to the possible cause. For example, an upstairs landing light when switched on during or after rain caused the relevant fuse to rupture. It was found that there was a roof leak and, on opening the ceiling rose, it was evident that polluted water had entered and caused a flashover between the phase and neutral and/or the protective conductor. After drying out, the circuit functioned normally.

A similar problem occurred in another house, but in this case there was no roof leak, but the upstairs lighting circuit MCB tripped occasionally when the lighting switch was operated. Again, the ceiling rose showed evidence of flashovers and it was found that the point was underneath a cold water tank in the loft. Condensation sometimes occurred on the exterior tank walls and base and dripped through the cable hole in the ceiling plaster into the ceiling rose.

In another case, the MCB controlling the downstairs socket outlet ring main tripped after the bath, on the first floor, had been used. There was a defective joint in the waste pipe under the bath and polluted water from it ran down a hollow partition wall nearby and entered a socket outlet box recessed in the partition, causing tracking across the insulation between the terminals at the back of the socket outlet and again triggering a flashover when wet.

Outages also occur when there is nothing switched on, during the silent hours for example. These are usually due to defective insulation vulnerable to a flashover triggered by HV or other transients which attain a higher potential when there is no load impedance to attenuate them.

Complaints about tripping out also occur where the whole installation is controlled by an RCCB. There is always some leakage from cooker, kettle and immersion heater elements which together with leakage from radio interference suppressors can often be enough to cause RCCB tripping, which is one reason why installation designers tend to restrict RCCBs to socket outlet circuits. There are often complaints about short lives of tungsten filament lamps. Where there is any substance in such complaints, it is usually due to an overhigh supply voltage. If this exceeds the statutory limits, the remedy is to get the supply company to put it right.

Warning

As a professional person, the engineer carrying out the checks has a duty to warn the tenant of any substantial dangers present and what he or she should do to be safe until the landlord rectifies them. Some dangers, such as exposed live conductors accessible to children, need immediate temporary measures such as making the circuit dead, or applying insulation or a barrier. Other items, such as the failure to apply equipotential bonding, are less of a threat and do not demand such immediate attention. Where the engineer considers that the dangers require it, he should advise the tenants' solicitor to seek a court injunction for immediate remedial attention.

The report

The report should state the qualifications and experience of the engineer either at the start or in an appendix as they will be called for in court if the case goes to trial. Then there should be a brief outline of the circumstances requiring the engineer's services. There should be a section of details of the installation, the defects found, and another for the results of the tests. Then comes a schedule of the BS 7671 infringements, quoting the number of each relevant Regulation. Note that unserviceable items which are not dangerous are not covered by the standard, which is concerned only with safety. Such items should be listed as a loss of amenity and are usually covered by the landlord and tenant legislation which requires the landlord to maintain the installation in a serviceable condition. In the conclusions should be listed the risks of electric shock and/or burn or fire to which the tenant has been exposed by the installation defects found due to the landlord's negligence. The final section consists of the work needed to remedy the defects. In cases where the installation is obsolete and where it would be unsatisfactory and uneconomical to repair, the engineer should say so and recommend a rewire.

Submitting the report

A solicitor to whom the report is submitted will usually require at least three copies of the report, one of which he will normally send to the landlord and request him to remedy the defects or face a court action. If the report has been properly prepared and it is evident to the landlord that the repairs are necessary, he will usually have them done. If there is any doubt about the remedial work, the solicitor will probably ask the report's author to revisit and to check it. In the event of the landlord's refusal to do the repairs and court action occurring, the engineer may be required to attend the court to give evidence as an expert witness. In this role, his or her responsibility is to the court, not to the client, and he or she has to assist the court by providing an unbiased opinion.

Index